GLASS

COLLINS ARCHAEOLOGY

GLASS

RUTH HURST VOSE

COLLINS
14 St James's Place, London

William Collins Sons & Co Ltd
London · Glasgow · Sydney · Auckland
Toronto · Johannesburg

First published 1980

ISBN 0 00 211379 1

Set in Bembo
Made and Printed in Great Britain by
William Collins Sons & Co Ltd Glasgow

EDITORS' FOREWORD

Additions to the large and growing number of books on archaeological subjects may seem at first sight hard to justify; there would appear to be a book to meet every need and answer every question. Yet this overprovision is more apparent than real. Although certain subjects, for instance Roman Britain, are quite well provided for, others have scarcely been touched. But more than that, archaeology is moving so fast on all fronts that the rapid changes within it make it very difficult for the ordinary reader to keep up. In the last twenty years or so there has been a considerable increase in the number of professional archaeologists who are generating a great deal of new knowledge and many new ideas which often cannot be quickly shared with a wider public. Threats to sites by advancing development in town centres, building on new land and road works, as well as from agriculture and forestry, have grown to terrifying proportions, and are being at any rate partially met by extensive rescue operations. The volume of ancient material of all kinds and periods has multiplied enormously, and its interpretion in the light of new knowledge and techniques has altered, in most cases radically, the accepted picture of the past.

There is thus a real danger of the general reader being left out of the new developments while the professionals are absorbed in gathering and processing clues. This series is intended for the reader who wishes to know what is happening in a given field. He may not be a trained archaeologist, although he may be attending courses in some aspect of the subject: he may want to know more about his locality, or about some particular aspect, problem or technique, or he may be merely generally interested in the roots of our civilization, and how knowledge about them is obtained. It is indeed vital to maintain links with our past, not only for inner enrichment, but for the fuller understanding of the present, which will inform and guide the shaping of our future.

The series presents books of moderate length, well illustrated, on various aspects of archaeology, special topics, regions, techniques and problems. They are written in straightforward language by experts in their fields who, from a deep study of their subjects, have something fresh and stimulating to say about them. They are essentially up-to-date and down-to-earth. They point to sites and museums to visit, and to books which will enable the reader to follow up points which intrigue him. Finally, they do not avoid controversy, because the editors are convinced that the public enjoys being taken to the very frontiers of knowledge.

CHERRY LAVELL
ERIC WOOD

Some Forthcoming Titles

HOUSES R. W. Brunskill
IRON AND STEEL D. W. Crossley
CHURCH ARCHAEOLOGY Ann Hamlin
THE FLOATING FRONTIER Mark Hassall
SOUTH WEST BRITAIN Susan Pearce
PREHISTORIC AND ROMAN AGRICULTURE Peter Reynolds
SETTLEMENT IN BRITAIN Christopher Taylor

Published Titles

VIKING AGE SCULPTURE Richard N. Bailey
THE ARCHAEOLOGY OF THE WELSH MARCHES S. C. Stanford
MATHEMATICS IN ARCHAEOLOGY Clive Orton

To Peter

ACKNOWLEDGEMENTS

My sincerest thanks go to Robert J. Charleston, former Keeper of Ceramics and Glass at the Victoria and Albert Museum, London, and President of the Glass Circle, who read through the manuscript of this book, offering many helpful suggestions for its improvement. Hilary Davies' meticulous work on the text is much appreciated.

I should also like to thank the following for their kind co-operation in putting this book together: Michael Archer, Deputy Keeper of the Department of Ceramics, Victoria and Albert Museum, London; John Atkinson, Conservation Officer with the North West Museum and Art Gallery Service; Derek W. Appleton, Editor, Ormskirk Advertiser Group; Pauline Beswick, Keeper of Antiquities at Sheffield City Museums; G. W. T. Bird, Pilkington Brothers Ltd, St Helens; Gavin Bowie and the Department of Finance for Northern Ireland; Robert H. Brill of the Corning Museum of Glass, New York; Dorothy Charlesworth, Department of the Environment, London; Barbara Clayton, Keeper of Archaeology, County Museum, Warwick; Neil Cossons, Director, Ironbridge Gorge Museum Trust; David W. Crossley, Department of Economic History, University of Sheffield; P. W. Elkin, Curator of Technology, City Museum and Art Gallery, Bristol; Geoffrey Emery of the Keele Archaeological Group; W. J. Ford, Library and Museum Service, Wiltshire; Charles R. Hajdamach, Keeper of Glass and Fine Art, Dudley Metropolitan Borough; Katherine F. Hartley, Director of the Mancetter excavations; J. Paul Hudson, Museum Curator, Jamestown, Virginia; Ivor Noël Hume, Director, Department of Archaeology, Colonial Williamsburg, Virginia; Journal of Post-Medieval Archaeology; G. H. Kenyon; Dominick Labino; Cherry Lavell, Council for British Archaeology; Lloyd R. Laing, Department of Medieval History, University of Liverpool; Frances Neale; P. W. Nelson, Group Publicity Manager, and H. A. Oakley of United Glass Ltd., Staines, Surrey; Penelope A. Pemberton, former Group Archivist, Pilkington Brothers Ltd., St Helens; Ada Polak; Ann Rigby, Librarian, Pilkington R. & D. Laboratories, Lathom; E. M. G. Robson, Information Officer, Glass Manufacturers' Federation; Gerald Shaw, Principal Technologist, Pilkington Brothers Ltd; Dinah Stobbs, Deputy Group Archivist, Pilkington Brothers Ltd; W. E. C. Stuart, Stuart and Sons Ltd, Stourbridge;

Trustees of the British Museum; Trustees of the Ryedale Folk Museum; Ray Wallwork, Coventry Archaeological Society; West Lancashire Library staff; Philip Whatmore, Honorary Treasurer of the Glass Circle; John Whitehead, Group Technical Communications Manager, Pilkington Brothers Ltd; David Wood, Group Public Relations Manager, Pilkington Brothers Ltd; Eric S. Wood.

Special thanks go to the Pilkington Glass Museum staff, Ian Burgoyne, Museum Manager, and Jennifer Metcalfe, former Administrative Assistant, who allowed me free access to the library, collections and photographic material. Barry Francis, Pilkington photographer, produced the majority of the excellent prints used in this book. Especial mention should be made of Denis Ashurst of Barnsley Archaeological Society, and Herbert W. Woodward of Brierley Hill, who both put themselves out on my behalf. Thanks go also to my old archaeological team at Pilkington: Freda J. Burke, Frank Bissette and Barry Francis, without whom much of this book would not have been written.

I am indebted to the many glass historians, archaeologists, manufacturers, artists, collectors, dealers and enthusiasts from all over the world from whom it has been my very real pleasure to learn and appreciate the history and techniques of glassmaking.

My apologies go to those people who provided me with information on excavated glass which, owing to limitations of space, I have been unable to use.

Marie Burke typed the manuscript.

My husband Jim never failed to encourage and assist me during the time it took to write this book.

R.H.V.

CONTENTS

PREFACE

Modern man takes glass for granted. Without it, there would be no windows for protection from the weather and admitting light into buildings. There would be no electric light bulbs to illuminate darkness, no lenses for spectacles, microscopes, telescopes and cameras. A life without television, glass containers and glass mirrors is now unthinkable; and how would we manage without the benefit of toughened and laminated glass windows for vehicles, and fibreglass for cars, boats, aeroplanes and houses?

And yet it is only in the last 150 years that man has been able to adopt a casual attitude to glass products, largely as a result of mechanized mass production. Before our technologically advanced era, the glassmaker was regarded in the popular imagination as akin to the alchemist – the maker of gold from the elements: his fiery furnaces dramatically transformed base materials into a translucent, sparkling substance which was valued by the rich and coveted by the poor. Glassmaking families guarded their formulae and production methods jealously, only passing them on from father to son, and this air of mystery continues today as large glassmaking firms file patent after patent to guard their glassmaking discoveries from competitors.

From very early times glassmakers have been honoured in various ways. Nearly two thousand years ago Roman glassmakers were given a special street in the better part of the city where they could practise their art. The Theodosian Code of 438 AD exempted glassmakers from taxation. During the Renaissance period, a noble person might marry into a Venetian glassmaking family with no loss of rank, and similarly, a French aristocrat might engage in glassmaking without upsetting his peers. In 1448 a charter was granted to Lorraine glassmakers allowing them the title of 'gentlemen glassmakers' and according them the right to rank equally with noblemen in such important matters as tax exemption. From Queen Elizabeth I's time,

British royalty and government have taken a keen interest in the prestigious and highly lucrative glassmaking trade – George Villiers, Duke of Buckingham (1628–87) being an outstanding example. British seventeenth-century parish registers in glassmaking areas in many cases show the word 'gentleman' after glassmakers' names, a certain sign of prestige in those socially rigid times.

This book aims to introduce the general reader to a fascinating craft whose history has ranged over four thousand years. It should also give a firm basis to any archaeologist wishing to tackle a glasshouse site. A short explanation of the composition and properties of glass is followed by an account of the origins and first techniques of glassmaking. The chapters on glassmaking in Europe from the Renaissance provide a background for the more specific history of British glass manufacture from the Bronze Age to Victorian times, and a discussion of the evolution of the British glasshouse. A description of the techniques of glasshouse excavation follows. Industrial archaeology has become increasingly popular in the past decade, and I have attempted to survey what is left of glasshouse structures and machinery no longer in use in an appendix on sites to visit.

Owing to the sheer magnitude of the subject it has not been possible to cover all aspects of the art of glassmaking. Window (flat) glass manufacture, including the art of stained glass, has been mentioned only briefly since it forms a large subject in its own right. The reader will find some selective guidance for further study of glass in the bibliography. A comprehensive checklist of recently published articles and books on glass is included in every volume of the *Journal of Glass Studies*, published by the Corning Museum of Glass, New York, since 1959.

PLATES

Unless otherwise stated, all plates are reproduced by courtesy of the Pilkington Glass Museum

FIGURES

MAPS

CHAPTER ONE

What is Glass?

The substance in your window panes is actually a flowing liquid. Delicate and lengthy experiments are needed to measure this flow; but they reveal that glass has the structure and all the properties of a liquid (*fig 1*).

A sheet of glass can be regarded as one large molecule, which explains why it is transparent: a ray of light passing through it hits only two optical boundaries, with a light loss of only eight to ten per cent. Glass does, however, refract light: a ray of light passing through a parallel-sided block of glass will emerge travelling parallel to its original path but displaced laterally from it (*fig 2*). It is this property which is used in lenses.

The light-dispersive quality of glass has been exploited by glass-cutters and engravers for more than two and a half thousand years. It was first explained by Isaac Newton, who realized that a prism of glass separates white light into its constituent colours, creating the brilliant rainbow effect we see in cut-glass vessels, and in the prisms of chandeliers (*fig 3*).

Although glass is commonly considered a fragile material, its strength in some respects is spectacular. A newly formed glass fibre will support weights of well over 70,000 kg/cm^2 (one kilogram per square centimetre is roughly fourteen pounds to the square inch).[1] This is twice as much as steel could support even in theory. The actual strength of ordinary glass is, however, accepted as only one hundredth of its theoretical strength owing to defects that lead to cracks. Because glass has the homogeneous structure of a liquid a crack, once started, will meet no obstruction to hinder its progress. Glass is weak in tension, though very strong in compression, and it is a poor conductor of heat, so sudden changes of temperature can cause dangerous stresses, as the skin of the glass becomes suddenly colder or hotter than its interior. Sudden heating, which puts the glass surface under compression, is less dangerous than sudden cooling, which puts the surface in tension.

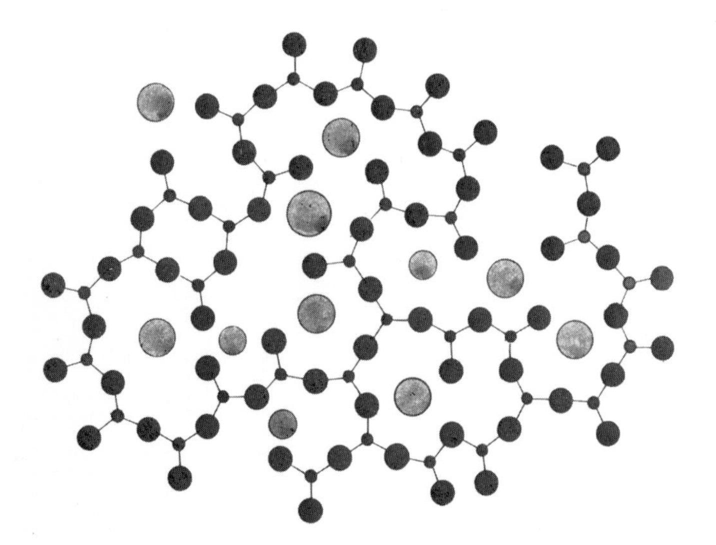

Fig 1 A small section of a two-dimensional representation of the structure of soda-silica glass. The large isolated grey rings are sodium (Na), the black rings with three arms are silicon (Si), and the larger black rings with one or two arms are oxygen (O). The bond between silicon and oxygen in glass is very strong, hence it is extremely difficult for this atomic structure to relinquish its liquid form to crystallize while cooling down.

When glass is heated it becomes soft and pliable. Ever since Roman times this ductile quality has been fully exploited by glassmakers. They learned to blow glass into bubbles, mould it into shape while blowing, add hot glass trails and blobs to its surface, and stretch and pinch it into shape.

Glass has a very high electrical resistance which has resulted in its modern use for electrical insulators. It is also dense and non-porous, which makes it ideal for containers and windows.

Although the majority of glasses are an invention of man, some glasses occur naturally and were certainly used by early peoples to produce primitive weapons. Volcanic glass, or obsidian, is formed by quickly chilled lava. It is black in colour, acidic in composition and like all glasses has a distinctive conchoidal, shell-like, fracture which can easily be trimmed to give a sharp cutting edge for making arrowheads, knife blades and other tools and ornaments. Obsidian occurs in many places in the Old World, such as the Greek island of Melos, and is

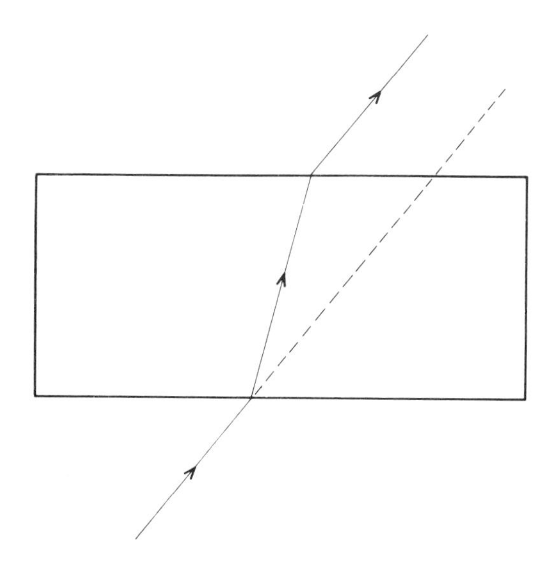

Fig 2 Light refraction in glass: a ray of light passes through a glass block and emerges parallel to its original path, but displaced laterally from it.

notably found in Yellowstone Park, U.S.A., on Mount Hecla in Iceland and the Lipari Islands in the Mediterranean. Other allied forms of natural glass include pitchstone, which contains more water than obsidian and is red or brown in colour. Techylyte volcanic glass occurs at the margins of basic lava flows, sills and dykes. Pelé's Hair is the name used to describe the golden brown fibres of volcanic glass formed from lava spray found in the Hawaiian Islands. Objects made from natural glass have been discovered on sites all over the world, and even in modern times the tribesmen of Africa and Australia have used the glass from bottles and even insulators on remote telephone and telegraph lines to make objects for their own use (*pl 1*).

Although glass is often referred to as 'crystal', its composition bears no relation to that of rock crystal, which is a mineral – a clear quartz. Since before the time of Christ glassmakers were probably attempting to imitate rock crystal when they made clear colourless glass; and the Italians referred to their clear glass as '*cristallo*' – which may partly explain why the term is used for glass today.

Most artificial glasses are and always have been made from a basic combination of sand, soda and lime. The soda (or another alkali, such as potash) acts as a 'flux' to bring the melting temperature of the silica or sand down from around 1700°C to about 1400–1500°C, and the limestone acts as a hardening agent. Modern glass technology is extremely complex, and the ingredients that go into making a glass today are very diverse. Besides the parent glass-former silicon dioxide, the oxides, sulphides, tellurides and selenides of substances like boron, germanium, vanadium, zirconium and arsenic are also used.

Fractional amounts of impurities in a glass 'mix' or combination of ingredients will cause the glass to be tinted and virtually opaque. The most common impurity encountered by glassmakers is iron oxide; a minute quantity of this can cause the glass to turn a variety of green and brown tints. Even today when sand quality is rigidly controlled, tints of green can easily be detected on the edges of panes of window glass.

Glass is altogether a most surprising material. It has been suggested that it should, as a super-cooled liquid, be considered a fourth state of matter, in addition to solids, gases and liquids.[2] No other material is so strong, yet so weak, so beautiful and yet so practical. Research into its nature and applications is continuing on a large scale, and recent developments include toughened, laminated and float window glass, glass ceramic ware, heat-and-light-sensitive glasses, and new building materials that combine fibreglass and cement. Despite the advent of competitors like plastics, the future of glass seems assured.

1. *Maloney 1967*, 30
2. *ibid* 17

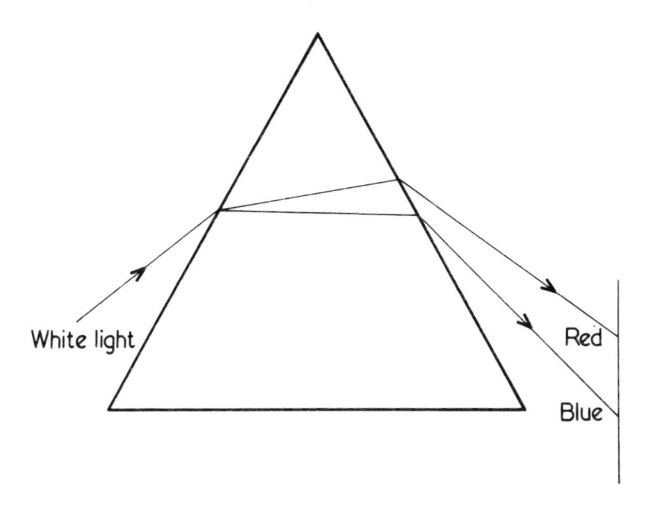

Fig 3 Light dispersion in glass: a ray of white light passing through a prism is split into its component rainbow colours.

Plate 1 An obsidian knife, thought to have been made by Californian Indians, and a glass arrowhead, chipped from a clear glass telephone insulator by Australian aborigines.

CHAPTER TWO

The Origins of Glassmaking

The question of when glass was first discovered and worked by man is one which has intrigued scholars for many centuries and has yet to be resolved. As early as the first century AD Pliny put forward his own theory of how the event came about. He describes how certain Phoenician (Syrian) merchants camping on the banks of the River Belus used cakes of natron (soda) they were transporting to support their cooking pots on the sands. When they woke in the morning they found the heat of their camp fire had fused the sand and soda together, forming glass. Certainly, Syrian glassmakers did use the sands of the Belus to make glass, but Pliny's story cannot possibly be true: a camp fire could not generate enough heat to fuse sand and soda. Tests show that the earliest known glasses were produced at temperatures around 1060°C.

The sheer impracticability of Pliny's story is not the only reason for dismissing it. Archaeologists have found glass on sites dating to the third millennium BC – more than 2000 years before Pliny lived – for example at Eridu on the Euphrates delta. These earliest glass discoveries consist of small objects, mostly beads.

Glass has certainly been used as a glaze on pottery for thousands of years; and a recipe for a glaze found on a tablet near Tell' Umar on the Tigris, dating to the seventeenth century BC, has given us our earliest information on glassmaking techniques.

Fragments of the first known glass vessels have been found on Western Asiatic sites such as Atchana/Alalakh, Nuzu, Assur and Tell al Rimah dating to the late sixteenth and early fifteenth centuries BC.[1] The oldest complete glass vessels have been found in Egypt, dating from the reign of Tuthmosis III (1504–1450 BC).

W. M. Flinders Petrie excavated the earliest glassmaking complex known to us at Tel el Amarna (*ca* 1375–1358 BC) where open hearths were apparently used with shallow crucibles supported on refractory drums.[2]

Although we do not know precisely when and where glass was first made it is now generally accepted that man has been making glass for over 4000 years, and that it was probably first produced in Western Asia.

Raw materials

Ancient Egypt abounded in the basic raw materials needed for glassmaking. Unlimited supplies of ordinary quartz sand and calcareous sand were available. Soda (natron) could be easily obtained from the vast desert deposits at Wadi Natrun, a depression in the Libyan desert some forty miles to the west of Cairo, and from El Kab in Upper Egypt, both mentioned in ancient Egyptian records. Limestone was available in unlimited quantities in the limestone hills bordering the Nile for about 500 miles from Cairo to beyond Esna. Dolomitic limestone and dolomite exist naturally in Egypt, and lime could also be obtained by burning sea shells. Alkaline salts forming the crude natron mentioned by Pliny could be leached out of the soil by seepage from the Nile; the dilute solution of salts could collect in depressions in the form of shallow ponds, and evaporation during the hot months would leave a deposit of alkaline salts at the bottom or around the shore.

A series of clay cuneiform tablets found in the Royal Library of Ashurbanipal (668–626 BC) at Nineveh gives detailed directions for preparing glass. Although these Assyrian tablets were written in the seventh century BC, it is thought on the basis of literary style that they could be copies of texts written some centuries earlier. The tablets mention the use of the soda-containing plant, salicornia, which can be found in most areas of the Near East, as an alkaline flux. In some places the Mesopotamian tablets specify 'ground red shells from the sea' which could be a lime ingredient. Unfortunately, though a great many recipes are given, very few are complete. The tablets also seem to describe three types of furnace, but none in proper detail.

Analyses of ancient glasses have revealed that, with few exceptions, glass of the Ptolemaic and Roman periods was of a basic soda-lime type, similar in many ways to ordinary modern glass. Twelfth Dynasty glass (*ca* 2000–1800 BC) was found to be a soda-lime glass, but after the Eighteenth Dynasty, analyses have found the glass to contain far more magnesia. The alkaline content of some Egyptian glasses is so high that it cannot derive entirely from natron but must have been added, probably in the form of a vegetable ash – maybe from the salicornia or 'naga' plant.

Ancient Mesopotamian glasses have revealed a much higher ratio of magnesia than contemporary Egyptian glasses, and can properly be called soda-lime-magnesia glasses. Glass containing lead as a major component occurs first in Mesopotamia, at least as early as the sixth century BC; and lead glasses of various kinds were the common type made in ancient China.

Roman writers tell us exactly where some of their glassmakers obtained their raw ingredients. The mouth of the River Belus on the Syrian coast, and a seashore deposit near the mouth of the River Volturnus, north-west of Naples, are both mentioned as sources of glassmaking sand. The sand of the River Belus was used by local glassmakers for many centuries and was exported for foreign use as well.

Roman glasses were often high in alumina, which made them highly durable. Much Roman glass seen today in museums is virtually in its original state because of the aluminous materials contained in the major batch materials or produced from the corrosion of refractories.

After the disintegration of the Roman Empire in the west during the first half of the fifth century AD, it would appear that glassmakers continued for a time to use the same ingredients as their Roman predecessors. Presumably they could still obtain the same raw materials such as the soda-charged marine plant-ash from the Mediterranean countries, even though western Europe was steadily becoming more remote from her Romanic past. After 1000 AD the political upheavals following the spread of Islam helped to cut medieval glassmakers off from their normal sources of soda and forced them instead to use potash (potassium carbonate), which they could obtain locally from the ashes of bracken and woodland plants. Potash, being another alkali, had the same effect as soda in a glass mix: it acted as a flux to lower the melting temperature of the silica or sand.

Furnaces

The best-preserved remains of Roman glasshouses have been found at Eigelstein, near Cologne, revealing both circular and rectangular furnaces.[3] However a clay lamp from the first century AD gives the best indication of what a Roman glass furnace looked like, showing two glassworkers at a furnace of at least two tiers, the lower being a stoke hole.[4] Excavations in Israel suggest that tank furnaces (in which the furnace itself is used to hold the molten glass instead of individual crucibles) were in use in the Syrian area in Roman times.[5]

Plate 2 Small Roman flask simulating agate, probably made in Alexandria, first century AD. *Height 57 mm (2.25 in)*

COLOURED GLASS

The first known man-made glass was coloured by the addition of metallic oxides to the basic mix, and all artificial glass continued to be tinted in this way until about 800 BC when clear, almost colourless glass was first made. For over 1000 years coloured glass vied in beauty and value with the precious stones it deliberately emulated.

Archaeological and historical evidence bears witness to this fact. Glass was used extensively in decorating the burial goods of the Egyptian King Tutankhamun. The shapes and colouring of the earliest glass vessels produced by the Ancient Egyptians suggested their favourite jewels – turquoise, lapis lazuli, red jasper and porphyry. The 'Murrhine' bowls, mentioned by Pliny and other classical authors, could well have been made of patterned stones such as agate, sardonyx, or madrepore at first, though later glass imitations of these bowls were similarly called '*vasa murrhina*'. The Roman leader, Pompey, brought 2000 Murrhine bowls to Rome after his victory over the Persian king, Mithridates. Cleopatra's famous Murrhine collection was auctioned in Rome by the Emperor Augustus, and the bowls were sold for 70,000 sesterces (*pl 2*).

The decoration of glass during this early period before glass blowing was invented was very much in the hands of the hardstone engraver or lapidary. It would appear that the same craftsmen with the same tools worked on glass side by side with semi-precious stones, using the same style of engraved decoration for both.

Glassmakers continued to imitate semi-precious materials right through to the Roman period – at least 2000 years. Tomb finds in China have revealed that Chinese glassmakers from as early as the fourth century BC were making glass primarily in imitation of quartz, jade and other hard stones.

Detailed directions for preparing coloured glass are given on the tablets found in Ashurbanipal's library at Nineveh (see p 27). Colouring agents were exactly the same in all ancient glasses. Metallic oxides like iron and manganese may have been introduced into the glass accidentally through impure raw materials, but oxides like copper, cobalt, antimony and lead could only have been intentional additions. The same metallic oxides have been used by glassmakers to colour glass right up to the modern period.

Blue glass

Blue was the first colour favoured by glassmakers, who used the metal oxides of cobalt and copper to produce the varying shades found in ancient glass. The earliest known piece of cobalt blue glass comes from Eridu in Mesopotamia; it has been dated, at the very latest, to 2000 BC.[6] The glassmakers who produced it probably obtained the cobalt from Persia or the European countries bordering on the Mediterranean. Only a minute amount of cobalt oxide (0.05 per cent) is needed to give glass a deep blue colour.

Analyses of blue Egyptian glasses have shown that their makers used both cobalt and copper, though the cobalt blue glasses often contained small proportions of copper or manganese. Tests by Farnsworth and Ritchie on pieces of blue glass from Eighteenth Dynasty Egyptian sites, at least 600 years later than the Eridu glass, have revealed that Egyptian glassmakers used copper to make their bluish green glass, and cobalt to make pure blue or violet blue glass.[7]

The Romans continued to use copper and cobalt to colour their glasses blue, but analyses have shown they were equally aware that a blue colour could be produced by iron in its ferrous state. Iron is usually present in glass as a mixture of ferrous ions and ferric ions. The ferrous ions, which are strong absorbers of light in the red region of the light spectrum, tend to colour glass blue. However, the ferric ions, which are weak absorbers in the violet region, tend to colour glass yellow. (The combined effect of the two ions is green, the most familiar colour in glasses containing iron.) In order to produce a deep blue glass the glassmaker would increase the ferrous ions in a reducing (non-oxidizing) atmosphere.*

The tradition of colouring glass blue continued after the fall of the Roman Empire, dark blue glass being especially prevalent in western Europe in the seventh century AD. Blue glass occurs in both the Sassanian and Islamic periods, but it was not until the Renaissance in Europe that blue glass again became popular on a wide scale.

Red glass

There were two distinct types of opaque red glass in ancient times. In the earlier type, which continued in use, the colour was due to metallic copper in a colloidal state. A later type which was in use for a more limited period has much lead oxide in its composition, the colour and opacity being caused by small particles of cuprous oxide.

The process of manufacture was as follows: copper-containing glass was melted in a reducing atmosphere, causing particles of red cuprous oxide and/or copper itself to be precipitated in suspension throughout the glass. The addition of lead to the mixture would help the process, since lead increased the solubility of copper at high temperatures and caused this solubility to decrease rapidly as the temperature was lowered, thereby enabling more cuprous oxide to be precipitated. Most of the few early glasses that contain lead have this opaque red colour.[9] The earliest recorded example of a glass containing lead oxide as a major constituent was a sealing-wax red glass found at Nimrud dating between the eighth and the sixth centuries BC.

In the Roman period, Pliny mentions an opaque red glass called 'haematinum' and seems to imply it was manufactured locally. An analysis of red glass from one Italian site revealed it was a soda-lime-lead glass coloured with a high proportion of cuprous oxide. Glasses coloured red with cuprous oxide do not occur in Germano-Roman factories, possibly because glasses in that region were made more for

* A reducing atmosphere is one deficient in oxygen. A smoky, or reducing atmosphere can be produced simply by throwing a piece of green wood (wood being basically carbon) into molten glass.

Robert Brill has pointed out that early glassmakers were obviously aware of the difference between oxidizing and reducing atmospheres: the Mesopotamian recipes directed them to use either a smoky (reducing, non-oxidizing) fire or a smokeless (oxidizing) fire in their furnaces.[8]

utilitarian purposes, whereas glasses made in Egypt and Mesopotamia were more for ornament.

In the medieval period red glass assumed a new importance in the production of stained glass windows for churches. Glassmakers continued to use copper for rich reds, but manganese was also used to make a pale rose-red or pink glass.

White glass

Opaque white glass has been popular with glassmakers in so far as it can be used to imitate other materials, notably porcelain. White glass was made in ancient times, but was usually included with other colours to resemble semi-precious stones like marble, agate and onyx. Very occasionally it was used as the main colour for a vessel. Early glassmakers produced their white opaque glasses with a calcium-antimony compound. Antimony, the standard decolorant of antiquity, could also produce opacity in white, yellow, blue and green glasses, if handled differently by the glassmaker.

White glass retained its rather secondary position in Roman times, appearing more as a decoration than a colour in its own right. It was sometimes used to produce a rope pattern, usually on the edge of bowls. A thread of white glass was twisted and embedded in a clear or coloured glass. This is claimed to be the forerunner of the Venetian filigree glass which was invented about 1500 years later. The luxury glassmaking centre of Alexandria in Roman Egypt used opaque white glass in mosaic bowls and in agate, onyx and marbled glassware. Examples of opaque white Roman vessels with no other colour added do occur in the form of lidded boxes, vases and head flasks including Janus flasks. After its virtual disappearance during the Dark Ages the Venetians reinvented opaque white glass for the European markets.

Green glass

The green, blue-green and amber-green tints found in so much ancient glass are the result of iron impurities in the raw materials, rather than the glassmaker's conscious desire for these colours. Ancient green-tinted glasses were intended for more common usage since they appeared in such quantity, particularly from the Roman period. The production of this cruder green glass survived the disappearance of the Roman Empire, presumably since it required far less skill on the part of the glassmaker, and most Dark Age and medieval glass in western Europe has either green or amber tints.

When Mediterranean marine plant ash stopped being imported into western Europe after about 1000 AD, glassmakers took to the woodlands to find local sources of alkali to use as a flux. They used ashes of bracken or beechwood or other woodland plants to provide the potash for making simple, utilitarian glass objects. Iron and manganese in wood ash could colour glass green, blue, amber or pinkish purple. The glassmaker could control the colour to some extent, probably by varying the furnace temperature and/or the time and temperature of the melting process.[10] It is known that glass was being produced in this way in the forests of southern Germany as early as the thirteenth century AD. Green '*Waldglas*' or forest glass continued to be made in northern and western Europe, including Germany, Bohemia, France (where it was known as '*verre de fougère*') and Britain until as late as the seventeenth century.

It is impossible to say whether any glass was deliberately coloured green until the Venetian green glass of the Renaissance.

Black glass

From the second to the first century BC and later, iron oxide was used to make very dark or black glasses. It is worth noting that an excess of any oxide will colour glass so deeply that it looks black. Examples of black glass during the ancient period are comparatively rare, and presumably it was not generally popular.

Yellow glass

Opaque yellow glass has been produced by glassmakers since antiquity. The first dateable Egyptian vessels, which bear the cartouche of Tuthmosis III (1504–1450 BC), have opaque yellow trailing. Early glassmakers used iron in its ferric state (see p 30) to produce a yellow colouring. Ancient yellow glass was probably made opaque by the use of antimony. Relatively few yellow glass vessels were produced in Roman times. Medieval glassmakers coloured the surface of flat glass with a silver compound – a technique that was widely used for stained glass windows.

CLEAR COLOURLESS GLASS

From finds made in Turkey and Mesopotamia it is clear that glassmakers were attempting to produce a clear colourless material free of all impurities from at least as early as the eighth century BC. Clear, colourless glass was probably being produced in Persia, Greece and Mesopotamia from the fourth century BC, and Egypt's Alexandrian workshops are thought to have made clear, colourless bowls, cast and ground to shape, from the third century BC onwards.

Only a small number of vessels made of transparent glass has survived from antiquity, mostly bowls – the earliest known being the glasses found at Gordion in Turkey and Nimrud in Mesopotamia, dating from between the late eighth and the seventh centuries BC. These rare glasses are mostly of high quality and were probably among the glass luxury ware of their time.[11] They were usually made of a clear or greenish transparent glass, ground, cut and polished with a high degree of technical competence.

Glass can be clarified by antimony. If a green-tinted glass mix containing antimony is heated to a certain temperature it will become crystal clear. (Antimony oxidizes the iron that causes the green tint.) Now since antimony was a common ingredient in glassmaking from at least 2000 BC – it was used as an opacifier – it was inevitable that at some point glassmakers would stumble upon its clarifying power. Certainly this discovery had been made by Roman times, and it was Roman glassmakers who first produced transparent and nearly colourless glass in large quantities. Tests have proved that Roman colourless glass was produced at temperatures around 1100°C.

Antimony went out of fashion as a decolorant after the fourth century AD and was replaced by manganese. This may have been because whereas antimony probably had to be imported in the form of antimonite, a local ore found in the eastern part of the Empire, manganese was generally available in the local clays associated with flints. The Romans may also have been aware of the dangers of using antimony, a very toxic compound, which in an uncontrolled furnace atmosphere could be lethal. Manganese replaced antimony as a decolorant for ancient glassmakers from this time right through the Islamic centuries. When added to glass as manganese dioxide, it acts as an oxidizing agent; and by oxidizing the ferrous iron to ferric, it partly removes the greenish hues caused by iron impurities. It should be noted that ancient glassmakers also used manganese as a colorant, and in some valence states it can cause an amethyst colour. When clear glass has been exposed to the sun's rays for a very long time, the reoxidation of reduced manganese by the action of ultraviolet rays can give the glass these pinkish tints. This solarization effect is responsible for the violet colours of so-called 'desert' glasses.

With the onset of the Dark Ages in Europe, the tradition of making clear, colourless glass emulating rock crystal passed to the Near

Eastern countries. The production of a clear, almost colourless glassware was established in Persia and Mesopotamia during the great revival of the glass industry there during the ninth and tenth centuries AD. Although this Near Eastern glass still had slight tints of yellow or green, its quality was generally good, with few bubbles, and the glassmakers sought to overcome any defects in their metal by using facet, linear and relief carving (*pl 3*). Europe, however, had to wait for the rise of Venice during the Renaissance period before the production of clear, colourless glass was properly re-established in the West.

Plate 3 Slightly weathered bottle in pale green clear glass with linear and facet engraving; Persian, tenth century AD. *Height 216 mm (8.5 in)*

Plate 4 Core-formed *amphoriskos* in dark blue glass with opaque white combed and marvered-in threads; made in the eastern Mediterranean, second—first century BC. *Height 175 mm (6.88 in)*

EARLY GLASSMAKING TECHNIQUES

About 1500 years elapsed between the making of the first glass vessels and the discovery of glass-blowing. Glassmaking without glass-blowing seems almost inconceivable today, but several distinct techniques were developed during this early phase of the industry.

The core technique

The first glass vessels ever made were produced by the skilled and extremely laborious method now known as the 'core' technique. It has often been called the 'sand-core' technique owing to the long-standing belief that the core round which the vessel was formed was made of sand. Recent research has shown this to be wrong; the core was more likely to have been made of mud bonded with straw.[12]

Core vessels normally average between ten and twenty centimetres in height, though much larger ones have been found in Egyptian royal tombs. They are thought to have been used as containers for perfumes, cosmetics and unguents. The finest examples of early core vessels come from the Egyptian workshops of the Eighteenth and Nineteenth Dynasties, including the time when el Amarna, the new city founded by Akhenaten (1379–62 BC), was flourishing. The excavation of Tel el Amarna by Sir Flinders Petrie showed that glassmaking had been carried out there on a large scale. Although Egypt exported core vessels to Syria, the Levant and Cyprus, Mesopotamian core vessels are distinctive enough in shape to indicate quite separate workshops.

A gap of several centuries apparently occurred in the production of core vessels after about 1200 BC; they reappeared on a commercial scale in the seventh century BC in the Levant. Presumably during that intervening period the core technique was preserved in Mesopotamia and possibly Syria, until the Phoenicians revived the craft. Core vessels of the sixth and fifth centuries BC show new forms based on the Greek shapes of alabastron, oinochoe, amphoriskos and aryballos. A degeneration of these forms was evident in the fourth and third centuries BC until the second and first centuries BC, when new shapes based on late Hellenistic styles emerged (*pl 4*). Core vessels were found in Italy from the seventh century BC onwards.

The process of making a core vessel is thought to have been as follows. A modelled core, possibly made of mud bonded with straw, was fixed to a metal rod and covered with molten glass. When the core was sufficiently covered, the outer surface was marvered (smoothed) on a flat stone slab. Decoration might be added at this stage. When the glass had cooled, the rod could be removed, and the core picked out.

The American glass technologist, Dominick Labino, has successfully produced facsimiles of the ancient core vessels by trailing hot glass from a dipstick on to a preheated core. The trailing was done in the furnace over a crucible (glassmaking pot), only removing the vessel from the heat to marver the surface of the glass. Trailed decoration, handles and feet were then applied to the vessel. After the temperature had been raised above the annealing range,* the vessel was slipped off the metal rod into the annealing oven to cool slowly. When the vessel was cold the core was picked out.[13]

Mosaic glass

One of the earliest techniques of glassmaking consisted of building up a vessel from small sections of glass rods fused together on a core, to form a mosaic. From finds at Tell al Rimah and 'Aqar Quf in Mesopotamia and Marlik and Hasanlu in north-west Iran, it is certain

* The purpose of the annealing furnace (or 'lehr' as it is called in modern times) was to cool glasses carefully, the slower and steadier the better, to relieve any tensions in the glass material which might cause it to shatter.

that mosaic glass was being made as early as the fifteenth century BC. These early fragments of mosaic glass came from high beakers, usually with knobbed bases or shallow dishes. Most of them were formed from circular sections of monochrome glass rods of various colours, though polychrome rods were also used.

The mosaic glass technique first involved the making of glass rods, a process which has probably changed little over the centuries. A 'gob' or 'gather' of molten glass would be gathered from the furnace on an iron rod by a glassworker. A colleague would dip another rod into the gob, attaching the molten glass, and the two men would walk rapidly in opposite directions, stretching the glass as they went. Within a few moments the glass would harden, and the glassmakers could cut the stretched rod to the size required. Polychrome rods would be made by gathering a different colour of glass over the first gather and stretching them together (*fig 4*). After cutting thin sections from the glass rods, the next stage in the mosaic technique was to stick the glass sections with some sort of adhesive to a core the shape of the inside of the vessel. An outer mould was added to hold the sections in place while they were fused together in the furnace. Once the vessel had been annealed both surfaces were ground smooth.

The technique apparently continued to be used in western Asia during Assyrian and Achaemenian times, and it was no doubt the Asiatic workers who introduced the technique to Alexandria when glassmaking was established there around the fourth century BC. Mosaic glass was produced in quantity at Alexandria and later, Italy, down to the first century AD.

The Ptolemaic glassmakers at Alexandria were equally famous for their glass plaques in fused mosaic glass. The patterns were built up from sections of mosaic rod fused together in a kiln and then stretched, so that the detail of the design was extremely fine. An analogy to the technique can be made with modern sea-side rock sweets where the letters of 'Brighton' or 'Blackpool' run right through the sweet, and appear back-to-front at one end. The pattern in mosaic rods runs through the rod in exactly the same way, which helps to explain why 'half' faces are sometimes found amongst mosaic glass. A section cut from a glass rod which depicted a half portrait head was simply reversed to complete the full face. This simple device produced a balanced portrait, cutting labour costs by half. Occasionally full faces were made in a single section. Surviving fragments show that mosaic plates made entirely of flowers, blossoms, leaves and stems were made in Alexandria. Decorative tesserae cut from polychrome glass rods have been found in Arslan-Tash and Nimrud dating from the mid-

Fig 4 Stages in drawing out a tube of glass, from the Diderot and D'Alembert encyclopaedia. 1 to 4 show the gather of glass being roughly blown and worked into a hollow shape, and attached to another iron rod. In 5 the two workmen are forming a tube by rapidly walking away from each other; on the floor resting on supports are two tubes they have already made. 6 shows the final operation of cutting the tubes to the required size and packing them together.

ninth to the seventh century BC, and also in glass inlays decorating the shrine of Nectanebo II of Egypt (359–341 BC).

Mosaic beads, probably originating from Alexandria, were part of the mosaic glass technique. Hot sections of mosaic glass were caught up on a heated glass bead which would be ground and polished once it had cooled (*pl 5*). Mosaic beads could also be made by simply using sections of mosaic rod, ground and polished, and bored through the centre.

Egyptian craftsmen who had become skilled in the mosaic technique were attracted to the courts of the early Islamic rulers and are known to have produced mosaic work in the ninth century AD at the

Abbasid court at Samarra, where finds of mosaic glass, probably used as a wall decoration, were discovered by German archaeologists in 1912–14.[14] Mosaic glass disappeared with the collapse of the Roman Empire until it was revived by the Venetians in Renaissance times in the form of '*millefiori*' glass, using a different process to gain the same effect.

Grinding and cutting

Grinding and cutting semi-precious minerals was a well-established practice before man-made glass appeared, and it is easy to understand from this why the abrasive techniques were among the first to be used on glassware. Although 'cold-cutting', as it is more simply known, was used from the beginning of glassmaking, it was only in the eighth century BC that its use became common. Sometimes, though infreq-

Plate 5 Egyptian mosaic beads, third—first century BC

Plate 6 Cold-cut finger ring in pale green clear glass, from Cyprus, second—third century AD. *Diameter 38 mm (1.5 in)*

uently, the technique was used to form a complete vessel from a raw, i.e. uncast, piece of glass such as the Sargon vase in the British Museum. More often a vessel or object would have been roughly cast, and the 'cold-cutting' used to finish off the best, usually colourless, glassware in Assyrian, Achaemenian and Hellenistic times (*pl 6*).

Casting and pressing

Early glassmakers borrowed the idea of casting and pressing glass into closed and open moulds from potters and faience workers. The method was used to produce not only vessels of open shape, but solid objects such as statuettes, inlays and ornaments.

For objects which were to be viewed from only one side, powdered or molten glass would have been pressed into an open mould, with the pattern formed on the inside. Early moulds were probably made of clay, and it is also probable that the designs on them were impressed from a wooden or similar mould. Once the glass was in the mould it would be fired in a kiln, then cooled slowly to take out any strains. When cold, the glass object would have been taken from the mould and any additional decoration, perhaps in the form of incised work, could then be completed.

Objects modelled in the round, to be viewed from all sides, were cast in two-piece moulds. Vessels produced in this way were usually open ones, such as bowls and dishes, since these shapes lend themselves to the technique.

Casting and pressing glass into moulds continued into Roman times, even after the technique of blowing glass into moulds was discovered. The Roman 'pillar' moulded bowl, a fairly common shape in Roman glassware, was used by Frederic Schüler for experiments with the '*cire perdue*' (lost wax) method at the Corning Museum of Glass, New York (*pl* 7). He proved that such bowls, with their shallow, heavily ribbed form, could be produced by this process, leaving a surface without flaws and requiring little additional grinding and polishing. He maintained that the mould must be made of a ceramic material and the temperature to melt the glass must be fairly high, ranging between 800°C and 1000°C.[15]

GLASS-BLOWING AND MOULDING

The discovery that glass, when hot, could be blown like a soap bubble on the end of a hollow, metal blowing rod was to revolutionize glass manufacture. It is hardly surprising that this major technical breakthrough in glassmaking coincided with the establishment of the Roman Empire in the first century BC, which provided such an impetus to artists, craftsmen and engineers.

It is thought that glass-blowing was perfected as a technique in Phoenicia (Syria) in the first century BC. At about the same time glassmakers learned to blow glass into moulds to give shape to the glass bubble. Suddenly the way was open for the first 'mass produced' vessels in glass. Since its discovery, glass-blowing has continued to be used in every glassmaking country up to and including modern times. It meant that, for the first time, glassmakers were exploiting the unique manipulative and ductile properties of glass; and the material began to be treasured in its own right, rather than as an imitation of other precious substances.

The technique spread rapidly throughout the Roman world. Some Phoenician glassmakers set up factories in the West and were making mould-blown vessels in Italy by the first century AD. From there, they advanced to the west and north of Europe, and were exploiting their new technique in Gaul, Germany and the Alpine provinces by 40 to 50 AD. The first century AD saw the foundation of the famous factories at Cologne and in the second century the north Gallic and Belgic factories were in full production. By the end of the second century,

Plate 7 Pillar-moulded bowl in blue-green clear glass, found at Sittingbourne, Kent. Roman Empire, late first century AD. *Height 79.5 mm (3.13 in); Diameter 190.5 mm (7.5 in)*

Roman glass manufacture had reached its peak; but close connections between eastern and western workshops can easily be detected with directly related mould-blown and free-blown (blown without a mould) work found in eastern, western and central glassmaking centres.

Apart from the Syrian coast, Egypt was the other main glassmaking centre; but while the Syrian glassmakers developed their glass blowing prowess, expanding their activities to the west and north, the Egyptians continued to produce luxury glassware using the old techniques, specializing in cameo-cutting, mosaic work and other fine coloured glasses. The Egyptians did not adopt the blowing technique to produce cheaper ware until the second century AD at the earliest, as is demonstrated by finds on various sites, especially Karanis. It has been suggested that Egyptian glassmakers learned the Syrian blowing technique in the glasshouses around Rome and Campania, since it is probable that Egypt exported glassmakers as well as her glass to Italy.

A Roman glassmaker using the technique of mould-blowing probably operated in the following manner. Once a gather of hot glass

Plate 8 Reconstructions of Roman glassmaking moulds.

had been taken from the furnace on the end of a blowing iron, it was placed inside a mould. The mould would be made of wood or clay with the required design formed on its inner surface. The glassmaker blew down his iron, forcing the glass bubble into the shape of the mould. After slightly cooling, the moulded bubble (paraison) was released from the mould (moulds could be in two, three or more sections which could be separated), the blowpipe was cracked off, and when the vessel was finished it was put to cool slowly in an annealing kiln. In modern glassmaking factories, the foot-operated moulds are kept wet so that the steam caused by the hot bubble touching the mould forms a natural protection, ensuring a longer life for the mould. Almost certainly ancient glassmakers did exactly the same to protect their laboriously produced moulds (*pl 8*).

Amongst the first mould-blown vessels known to us are the small ones from Sidon in Syria, a famous glassmaking centre. Often they are stamped with the name of the maker and the word 'Sidon'. Some of the most outstanding mould-blown wares produced in Roman times were those made by Ennion, one of the first glassmakers to practise the convenient habit of signing his own vessels. He appears to have worked first at Sidon, then to have moved on to the Italian peninsula, where his products included jugs, cups, and two-handled vases, decorated with stylized plant forms and reeding in the same manner as 'samian' ware, the glossy red pottery of the time. Other signatures appear on glasses, slightly later than Ennion, the best-known names being Jason, Meges, Neikaios, Artas and Ariston.

Head flasks and Janus (two-headed) flasks were produced in many parts of the Empire, though Syrian and Palestinian head flasks are reputed to be better finished (the mould seams are supposed to be less obtrusive) than the ones produced in Gaul and the Rhineland. Although they had important religious significance in early Greek and Cretan rites, head flasks were used for amusement in Nero's Rome. A flask reputed to depict Nero's court fool, a shoemaker, is supposed to have inspired the German term for these objects – '*Schustergläser*' (shoemaker glasses). So-called 'Victory' cups were glasses used to commemorate events in early Imperial times and were moulded with designs of laurels and inscriptions (*pl 9*). It is thought that Ennion's

Plate 9 Strap-handled 'Victory' flask in yellowish-green clear glass, mould-blown, produced in Syria, first century AD. *Height 133.5 mm (5.25 in)*

north Italian shop set up a subsidiary glasshouse at Brugg (Vindonissa) in Switzerland, or Lyon in France, where they produced mould-blown cups decorated with scenes of the arena or circus, now called 'Circus beakers'. Most circus beakers come from the Alpine, Gallic and Rhenish provinces and Britain.[16]

A frequent find on western European, especially Gallic, sites is a class of greenish moulded bottles of cylindrical or barrel shape with moulded hoops at top and bottom. The makers' names are frequently moulded on the bases, the most famous being that of Frontinus, who apparently had a glassworks in Gaul at Boulogne or Amiens, flourishing in the third or fourth centuries AD. Coins of Constantine and other fourth-century AD emperors have been found buried with the bottles in Roman graves. More common finds on any Roman excavation are the plain mould-blown jars, jugs and short-necked bottles, usually in thick bluish-green or greenish glass found in cylindrical, square and polygonal section. They have strong, angular handles, often longitudinally ribbed, known as 'celery' or 'strap' handles, and they appear to date from the first and second centuries AD.

All Dark Age glass was blown, but after about 400 AD there was a great reduction in the variety of shapes and decorative techniques used by the glassmakers. Besides simple jars and beakers, the most familiar shapes of this period were the claw, bag-shaped and cone beakers, palm cups and the drinking horn.

Mould-blown bottles and jugs bearing Jewish and Early Christian symbols form a distinctive group, now dated between 578 and 636 AD, and were probably all made at the same workshop in Jerusalem.[17] They are hexagonal and octagonal vessels in blue, green and purple, but more commonly, brown glass, with moulded decoration in intaglio, probably produced from metal moulds (*pl 10*).

Moulded decoration was used in the Near East in Islamic times, the patterns usually covering the whole body of the vessel. Similar vessels were made continuously for several centuries. Moulded sprinkler bottles in heavy green glass were a common product of the twelfth century, sometimes decorated with a vermicular collar, or in the earlier types, with a little bulb or bulbs in the cylindrical neck (*pl 11*).

DECORATION: BY THE GLASSMAKER

As soon as the first glass objects were made, glassmakers sought to embellish their creations by ornament. An intimate understanding of his medium, a craftsman's dexterity and an artist's eye were required

Plate 10 Mould-blown jug in clear amber glass with weathering, bearing early Christian symbols, Palestine, 578–636 AD. *Height 140 mm (5.5 in)*

of the glassmaker to decorate his wares successfully. Speed was essential: the glassmaker would have only a couple of minutes to gather and form his vessel, take another gather of hot glass for the applied decoration, and place the completed vessel in the annealing kiln to cool.

Trailing

The earliest known glass vessels, the little perfume bottles of Mesopotamia and Egypt produced by the core technique, are almost without exception decorated with glass trailing. The glassmakers would take a small gather of coloured glass from the furnace, contrasting with the body colour of the vessel, and trail the hot glass on to the surface of the equally hot vessel (any difference in temperature would set up strains in the glass, causing it to break). A pointed instrument was probably pulled over the surface of the hot glass to produce the various combed effects. The threaded decoration, still proud of the surface of the heated vessel, would be marvered or smoothed level with the surface on a flat stone slab or 'marver'. The trailing on core vessels is mostly in brilliant contrast to the vessel colour, bright yellows, brick reds, rich greens and ambers, and light blues being very popular.

Plate 11 Islamic sprinkler bottle in clear green glass with weathering, mould-blown, twelfth century AD. *Height 254 mm (10 in)*

During the Roman period, when the making of core vessels virtually ceased, the art of marvering trails of hot glass into a vessel continued on a limited scale on blown vessels. With the break-up of the Empire, the technique lived on in western Europe, where vessels were produced with threads of contrasted colour for decoration, looped up or combed and sometimes worked flush into the surface of the vessel. Gaul, the Rhineland and Spain produced glass of post-Roman date with this 'feathered' decoration. Islamic glass, irregular and almost shapeless in appearance, was sometimes decorated with combed and marvered white threading.

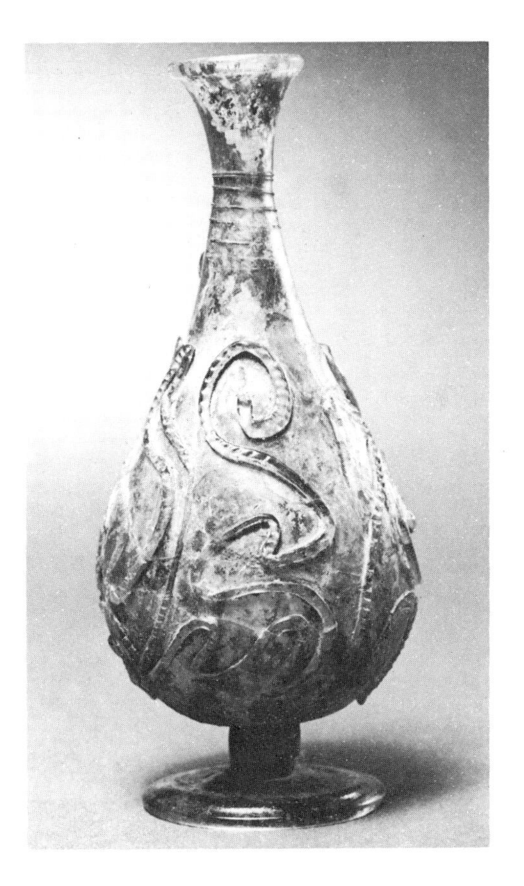

Plate 12 Clear green glass bottle, slightly weathered, with applied 'snake trails', Roman Empire, second—early fourth century AD. *Height 181 mm (7.13 in)*

Plate 13 Syrian 'dromedary' flask in clear yellowish glass, heavily weathered, sixth—eighth century AD. *Height 120 mm (4.7 in)*

Trailed decoration which stands proud of the main body of a vessel (i.e. not marvered in) was particularly favoured by Roman glassmakers. Glass with threading, simple or intricate, has been found on sites throughout the ancient empire. The so-called 'snake thread' or 'snake trailed' decoration distinguishes a particular group within this type. The threads, usually with a notched design, were set in irregular winding patterns, sometimes in spirals and occasionally with ivy motifs. Snake thread vessels with opaque white, yellow and blue trailing were produced mainly in Cologne glasshouses in the second century AD to the beginning of the fourth century. They were also produced in eastern workshops in the same period, usually with self-coloured trails (*pl 12*). Glass from the latter years of the Roman Empire is wild and irregular in form. Trailed decoration became extravagant towards the third century AD with elaborately tooled handles applied to the vessel and threads of glass dribbled over the surface. The next century was to see Roman glassmakers reaching the height of technical daring with threaded decoration creating tiers of tall loops for handles, and geometric lacings that stood away from the main body of the vessel, sometimes almost hiding the vessel beneath them. The later Roman glassmakers were not concerned with exact proportions, nor, apparently, with the quality of their metal, which tended to be impure and full of bubbles. Yet their glasses had freedom and grace of form, and great play was made with the trailing-on of handles, such as the so-called 'strap' and 'chain' handles (see p 46).

In Merovingian or Frankish times (fifth to eighth centuries AD) the only Roman decorative techniques to survive in the west were trailing and mould blowing. The Dark Age glassmakers could trail glass with as much dexterity as their Roman predecessors on their glass cone beakers, drinking horns and other simple vessels, all characterized by the absence of a practicable foot. Trailed decoration continued in popularity in western Europe into Renaissance times. In the Near East, from the sixth to the eighth centuries AD, free-standing threads of glass were used to create the open trellises of glass around the so-called 'dromedary flasks'. Probably originating from Syria, these vessels are in the shape of quadrupeds bearing vases on their backs, surrounded by trellis work built up from zig-zags of mainly self-coloured glasses (*pl 13*).

Blobs and prunts

The technique of dropping hot blobs of glass on to a heated vessel for decoration was fully developed by the Roman glassmakers. Hollow-blown prunts were used to create dolphin shapes to decorate glasses, and in the Rhineland, glasses ornamented by all kinds of sea creatures and snails are generally termed 'shell' or 'dolphin' beakers.

With the fall of Rome, the Germanic love for sturdy blobbed decoration soon revealed itself. It is thought that the blobs had some religious significance and were put on the glass to ward off evil spirits, but, practically, the blobs gave a sure grip to the hands of a people who regarded heavy drinking as a social obligation. A development of blobbed decoration occurred with the idea of re-blowing a heated vessel, once the hot blobs were applied, so that the blobs blew out further than the cooler walls of the vessel. So-called 'claw beakers' (German '*Rüsselbecher*'), originating from the Seine-Rhine area in the fifth century AD, were made in this way, the blown blob being caught by a small metal tool, and drawn to its final, claw-like shape.

A cruder form of applied decoration used by the Romans was the marvering-in of small, normally opaque, white, blue or red pieces of glass into the body of the vessel. Fragments of glass would be caught up and marvered into a gather of glass of contrasting colour, and a vessel formed from it. By working the gather on the marver the fragmented glass would be brought level with the surface of the gather, so that a smooth surface was achieved when the vessel was blown.

DECORATION: BY THE DECORATOR

The expertise of the decorator, which owes little to glassmaking technology, can be seen in the various forms of enamelling, gilding, cold painting, cutting and engraving used on glasses throughout the ages. One of the earliest glass vessels which bears the name of the Egyptian Pharaoh Tuthmosis III is decorated with powdered glass, fired on in the manner of enamel, enamelling being an art quite distinct from that of glassmaking. Incised decoration was found on a turquoise

Plate 14 Enamelled and gilded beaker in pale straw-tinted clear glass, slightly weathered, Islamic, *ca* 1170–1270. *Height 121 mm (4.75 in)*

blue opaque glass goblet, which bore the name of the same Pharaoh, and on other pieces of dynastic glass.

Enamelling

Enamelling means using a low-firing glass (one which melts at a low temperature) crushed to a powder and painted on to the surface of a vessel with a bonding agent. The powdered glass (enamel) will fuse to the vessel when placed in a small furnace at a low enough temperature to prevent the vessel warping or sagging in the heat.

Enamelling was practised in Egypt at the time of the Eighteenth Dynasty but apparently nowhere else. The technique recurred in a more advanced, sometimes magnificent, form during the Roman period. Motifs used by Roman enamellers included birds, animals, vines, pygmies, hunting scenes and figural subjects. There is no record of the exact process used by Roman enamellers, but from Robert Charleston's research it has been established that there was a trade in cakes or ingots of glass enamel for the use of less specialized glassmakers from earliest times, the most popular colours being turquoise, white and sealing-wax red.[18] The monk Theophilus, who wrote *Schedula Diversarum Artium* (*Treatise on Divers Arts*) probably in the first half of the twelfth century, devoted a section of his work to 'Glass Goblets which the Greeks embellish with Gold and Silver' and included a description of enamelling.

> Then they take the white, red and green glass, which are used for enamels, and carefully grind each one separately with water on a porphyry stone. With them, they paint small flowers and scrolls and other small things they want, in varied work, between the circles and scrolls, and a border round the rim. This, being moderately thick, they fire in a furnace in the above way . . .

Byzantine glasses of the type described by Theophilus have been recognized, dating from the eleventh and twelfth centuries.

A high point in the history of enamelling was reached in the thirteenth and fourteenth centuries in the Near East, especially in Syria. The great Islamic mosque lamps are the best-known Syrian products, many exported to adorn Egyptian mosques. The glass had greenish or brownish tints, but this was masked by the fine enamelling and gilding which virtually covered the vessel. Besides the mosque lamps, the Syrian enamellers decorated other objects such as sprinklers, globes, footed bowls, beakers and long-necked bottles. Like the Romans, the Islamic glassmakers and decorators left no written record of their craft (*pl 14*).

Plate 15 Graeco-Roman gold sandwich glass bowl, third—second century BC. *Diameter 105 mm (4.12 in)*

Cold painting

Cold-painted decoration is easy to distinguish from enamelling since it easily flakes off the glass, and is rarely found in good condition. Substances such as lacquer, varnish and oil pigments have been used to decorate glass, but, not being fired on to the glass, they rub off easily and must be considered an inferior form of decoration. Examples of Roman glass with unfired painting have been discovered.

Lustre painting

Lustre painting describes a somewhat mysterious form of decoration which was neither enamelling nor cold painting, but did involve a film of pigment being fired on to the glass. Depending on the firing, the film became more or less lustrous and smooth to the touch. The technique originated in Egypt, either in the late Byzantine or early Islamic period, and was possibly also known to Syrian glassmakers. Exactly how lustre painting was done has never been discovered, though there is good reason to connect it with the technique of lustre painting on tin-glazed pottery which was practised first in Mesopotamia in the ninth century AD.[19] The lustre produced was reddish-brown in colour, painted and fired on to a colourless glass surface, possibly slightly tinted with green through impurities.

Gilding

Gold was already being used by glassmakers by the beginning of the first millennium AD in the production of mosaic objects such as bowls, dishes and pyxides (small lidded containers) and also in the gold-band 'alabastra'. Still in the province of the glassmaker was the ancient technique now called 'gold sandwich glass' which involved the trapping of a gold design between two layers of glass. Some of the earliest examples of this technique were found at Canosa, Apulia, Italy, dating to the third century BC, but it reached its most prolific period in approximately the fourth century AD (*pl 15*). The majority of third and fourth-century AD gold sandwich glasses were found in the catacombs around Rome, embedded in the plaster of the '*loculi*', and were presumably put there by relatives of the dead. These early Christian '*fondi d'oro*' (another term used to describe gold sandwich glasses) were usually the circular bases of shallow bowls or dishes decorated by the technique. The sides of the bowls or dishes are almost always broken away, leaving only a few, tell-tale fragments.

Gold leaf, cut to the desired design, was put inside the base ring of the glass, or it was engraved with a point once it was in position, and occasionally painted decoration was added. This was covered with a further, closely fitting and protective layer of glass, which was fused or cemented to the base of the bowl. This outer layer of glass was sometimes coloured, but more often both layers were clear and colourless. The drawback to the technique, which sought to avoid the problem of wear on surface gilding, was that the glassmaker/decorator was not always successful in avoiding trapping air bubbles between the two layers, which could disfigure the design. The subjects depicted on these medallion-like embellishments were taken from Jewish and Christian symbolism and Biblical history, but pagan motifs were also used, such as scenes from games and classical mythology, and there are dedications to circus heroes, as well as to saints. Almost certainly, these gold sandwich glasses were manufactured at a workshop in, or near, Rome. The name of a fourth-century AD bishop, Damasus, has been found on several fragments of this glass, and since no more burials took place in the catacombs after about 410 AD, this would seem to be its latest date.

This method of gilding is believed to have been used in Rhenish glasshouses in Roman times, with fragments in a different style being found at Cologne. The same technique was used on a smaller scale to decorate coloured blobs of glass, dropped on to a colourless bowl. The gold decoration was placed between the inner faces of the blobs and the outer surface of the colourless vessel, creating a rich effect. A series of Byzantine tiles dating somewhere between the sixth and the twelfth centuries AD revealed a modification of the technique, with rectangular patches of gold leaf covered with a film of colourless glass for protection.

Roman glassmakers undoubtedly used gilding on the outer surface of the glass, often in conjunction with painting or enamelling, but the ravages of twenty centuries have left little evidence of it. Few gilded vessels have survived unscathed, since few glassmakers surmounted the problems of getting the gold to stick to the glass and retain its brilliance. The art of gilding glass disappeared in the West amongst the upheavals following the collapse of the Roman Empire, never to be revived until the Renaissance. But it was practised in Byzantium in the eleventh and twelfth centuries, and achieved great popularity in the Near East by the twelfth century when Egyptian craftsmen took service at the courts of the rulers of Syria and north-west Mesopotamia, taking the art of gilding with them (*pl 14*, p 52). The thirteenth and fourteenth centuries saw the next high point in the history of gilding, combined with enamelling, in the great Islamic mosque lamps and other vessels produced in the Near East, particularly in Syria (see p 53).

Cutting and engraving

Before man started to make glass, hieroglyphic inscriptions had been engraved on hard stone vessels, and as early as the sixteenth century BC glass was being engraved in Egypt. Glass vessels with incised inscriptions, such as that found by Flinders Petrie at Tel el Amarna, have survived from Egypt's Eighteenth Dynasty. According to Charleston's research, the engraving on these early vessels was probably produced with a pointed instrument, and it is only in the later Egyptian period that there is evidence of wheel-engraving on the glass, probably using a bow-lathe. Decorative motifs first produced by moulding processes were sharpened up by rotary abrasion on glass vessels in the general Mesopotamia-Assyria-Asia Minor region, notably on a large family of shallow bowls with radiating petal motifs, the earliest dating not later than 700 BC. Abrasive powder had to be fed on to the grinding or cutting wheel to abrade the glass, and it is probable that emery, which was known in Egypt from at least as early as the Eighteenth Dynasty, was used as an abrasive in ancient times. As Charleston notes, the Egyptians are known to have used tubular metal drills in cutting granite, and it was probably by some such means that vessels were hollowed out at this time. Although it is not known what type of abrading equipment was used by Roman glass engravers and cutters, it has been suggested that an all-purpose tool may have been adapted as lathe, drill or engraving wheel as needed.[20]

Facet-cut and 'diamond' point engraved bowls, probably from Egypt, appeared in the first and second century AD. They were made in clear, colourless glass, and decorated with mythological and genre scenes in facet-cutting with 'diamond' point engraving for the details, although a diamond was probably not in fact used. Large, broad wheels were necessary to produce the wider abrasions in facet-cutting. The writer Pliny mentioned some of the abrasives used for grinding in Roman times, such as 'sand of Naxos' for emery, 'sands' from India, Egypt and Nubia, and certain stones from Armenia and Cyprus. Theophrastus mentioned pumice and also emery in his *History of the Stones*. Pliny spoke of Thebaic stone from Egypt and pumice used for the final polishing of marbles. Facet-cut cups found in England and Cyprus from the Roman period are thought to originate from western Syria or Egypt, and, in the later second century AD, from eastern Syria.

The quality of Alexandrian engraved work deteriorated during the third and fourth centuries AD, but in the west a great quantity of figured, engraved and cut glass began to be produced, especially in and around Italy. Figured decoration was not entirely abandoned in the eastern workshops, but in general they concentrated on easier designs, sometimes geometric and rather rough in execution. Wheel-engraving was used to imitate facet-cutting, with designs of curved lines, circles and ovals. Many of the vessels of this type found in the West must have been imports from the west Syrian-Egyptian workshops. Roman '*diatreta*' glasses are counted amongst the world's most outstanding examples of the art of the abraded decoration. From experiments carried out by Fritz Schäfer in Germany and Barbini and Fuga in Italy, it is now virtually certain that these 'cage' cups were produced solely by means of abrasion.[21] Diatreta were the finest products of the Cologne and Trier glassmakers in the fourth century AD, made with a 'cage' of decoration carved from a solid blank, often including an inscription. Examples, often with casings in different colours, have been found in the Rhineland, on the Danube, in northern Italy and in Greece.

Cameo glass was another engraving technique developed by Roman glass decorators which combined both the heavier wheel-cutting and the lighter wheel-engraving. Cameo glass was certainly made in the Alexandrian workshops of Egypt in Roman times, particularly between the first century BC and the first century AD, when the technique had reached such perfection that masterpieces like the Portland Vase at the British Museum were being produced. Cameo glass made a brief revival after the fall of the Empire in Egypt and Persia in the ninth and tenth centuries AD. It formed part of the school of relief-cutting which flourished in Persia and probably Mesopotamia in the ninth and tenth centuries AD.

Cutting and engraving required good quality glassware, which was the main reason for the disappearance of abrasive techniques after the decline of the Roman Empire in the West. In the East, the technique never ceased, and glasses with abraded decoration can be found through to Islamic times. Baghdad and Basra were noted for cut glass from the ninth century AD, and the relief-cutting from Persia and probably Mesopotamia in the ninth and tenth centuries AD was not rivalled until the seventeenth century.

1. *Harden 1968*, 11
2. *Petrie 1894*
3. *Doppelfeld 1965*
4. *Abramić 1959,* 149ff
5. *Weinberg 1968,* 49–50
6. *Garner 1956, 147–9*
7. *Farnsworth and Ritchie 1938*
8. *Brill 1963*
9. *ibid*
10. *Newton 1978,* 59–60
11. *Saldern 1959,* 23–49
12. *Harden 1968*, 12
13. *Labino 1966,* 124–7
14. *Lamm 1928*
15. *Schüler 1959,* 47–52
16. *Painter 1968*, 36–40
17. *Barag 1970, 1971*
18. *Charleston 1963,* 58–60
19. *Pinder-Wilson 1968*, 101
20. *Charleston 1964,* 87
21. *Schäfer and Zecchin 1968,* 176–9

CHAPTER THREE

Glassmaking on the Continent: Middle Ages to 1700

Archaeological and documentary sources have provided us with a clear picture of glassmaking methods in Europe since the Middle Ages. Glasshouses have been excavated in the main centres of the medieval industry, in Italy, Bohemia, France, Germany, Denmark and Sweden.

It was probably during the first half of the twelfth century that the Westphalian Benedictine monk, Theophilus, published his important treatise 'on divers arts', *Schedula diversarum artium,* Book II of which is devoted to glassmaking. In it he describes the furnaces, crucibles, tools and ingredients used by contemporary glassmakers.

A great deal of our technical knowledge of medieval glassmaking derives from Antonio Neri, who first committed to paper the secrets of glassmaking '*à la façon de Venise*'. His book, entitled *L'Arte Vetraria*, was largely based on his experience in Antwerp glassworks founded by the Venetians. It was published in 1612, and became the standard book on Italian glassmaking.

Other major documentary sources include the writings of glassmakers like Johann Kunckel, who used Neri's work in writing his own *Ars Vitraria Experimentalis*, and the Swedish priest Peder Månsson, who on his travels abroad observed different glassmaking methods being used from the ones he was accustomed to at home.

Theophilus' book indicates that vessel glass was relatively rare in twelfth-century Europe, although he does describe how glass bottles were blown. Window glass was however beginning to be produced on a large scale at this time, for use particularly in churches and royal residences. There were two principle types of window glass: crown and cylinder. Crown glass was made by blowing a gather of glass into a bubble, attaching a pontil rod to the side opposite the blowing iron (which was then broken off), and spinning the rod until the open-ended bubble 'flashed' open, by centrifugal force, into a flat disc or 'crown'. This was then cut up into small panels, one of which would contain the familiar 'bull's eye' or centre piece, the mark left by the pontil rod. This technique came to Europe from the East, where it had been known since Roman times. Cylinder glass was made by blowing hot glass into a cylinder shape which was then slit along its length, flattened, and cut to size.

The chief centres of window glass production in western Europe were Burgundy, Lorraine and the Rhineland, which specialized in cylinder glass, and Normandy, which specialized in crown glass.

By the thirteenth century window glass was in use in other buildings besides churches and palaces, and by the beginning of the sixteenth century it was common in private houses of any pretension.

Medieval glassmaking was a seasonal business: we know that Venetian glasshouses worked only from January to August, and glasshouses in the Spessart Forest produced glass from Easter until Martinmas, in November. Furnaces were fuelled with wood: in southern Europe this would often be alder and willow, and in the north, in Lorraine and Normandy, beech and oak were used.[1] Once a furnace was lit, it was kept burning day and night.

The glasshouse consisted of a main furnace with several subsidiary kilns around it. It had been obvious to glassmakers from very early times that it was more economical to arrange the subsidiary kilns so that they could all benefit from the central source of heat. The subsidiary kilns were used for annealing (see p 37), 'fritting' (a preliminary melting process in which the raw materials were roughly fused together, then broken into lumps either for immediate use in the main furnace, or to be stored for future use) and preheating the clay crucibles in which the glass was actually made.

The making of crucibles was as important to glassmaking as the working of glass itself. At least one member of a glassmaking group would have had to be a skilled potmaker. Molten glass is a highly corrosive liquid, and whatever it comes into contact with has to be refractory – able to withstand high temperatures without fusion or decomposition. Early potmakers learned by trial and error that certain clays made better refractories than others. We now know scientifically that the most suitable clays are those with a high alumina/silica ratio.

It is reasonable to assume that the craft of potmaking has changed little through the centuries. The raw clay was allowed to stand for several months and was trodden with bare feet to remove any air bubbles and to compress it into a good working consistency. The base of the pot was formed first from rolls of clay, about ten centimetres long and five centimetres in diameter, thrown on to a circular board and kneaded into the required thickness by the potmaker. The pot base would then be removed from the board by means of wire-cutting or a band saw, the board having first been turned over.

After the base had been left to stand for a time, a rim would be worked up on the edge upon which the sides would be built, again from rolls of clay built upwards in sections, each being allowed to dry for a time before the next layer was added. The potmaker worked the clay all the time he handled it, to eliminate the possibility of any air bubbles being left in it which might expand on being heated and break the pot. Up to the invention of the English covered pot in the seventeenth or eighteenth century, all glassmaking pots were open, and came in a variety of shapes including barrel, bucket, slightly flared, and waisted with a flared rim.

The potmaker's skill was particularly vital in the firing process. The effect of the corrosiveness of molten glass on the pot could be much reduced if the pot was porous – and the porosity of the pot depended on the nature of the clay and the way it was fired. The quality of the clay could also directly affect the quality of the glass, since impurities in the clay would be absorbed into the metal as it gradually ate its way through the refractory.

After firing the pots were left for some months to dry out thoroughly and 'cure'. The pots were preheated in a separate 'pot-arch', and then transferred to the main furnace – an operation now known as 'pot-setting', and always an unpopular job. An even more unpopular task was clearing up the mess if a pot did collapse or break with the heat, depositing its molten contents over the interior of the furnace.

R. J. Charleston has proved that, during the later Middle Ages, two divergent glassmaking traditions developed in Europe: that of the north, including Germany, France, Belgium, Britain and Bohemia, and that of the south, mainly in Italy.[2] They used different types of furnace, different raw materials, and produced different kinds of glassware.

Furnaces

The furnaces used in northern Europe were generally rectangular in plan with the various compartments adjacent to and on the same level as the main chamber. They shared their heat by means of either a common fire channel running through the structure, or linnet holes which transmitted heat from the main furnace to the subsidiary ones. In his *Schedula Diversarum Artium* (see p 59), Theophilus described a rectangular furnace, but it is evident that there were considerable variations on this furnace pattern from the descriptions in Chapters VII and VIII of *De Coloribus et Artibus Romanorum*, a tenth-century text, probably added to in the twelfth and thirteenth centuries, and attributed to a certain Eraclius. He describes an oven composed of three chambers of unequal size, the large main working furnace being in the centre, flanked by a second fritting chamber, and a third pot-firing compartment.

Fig 5 Fifteenth-century drawing of a glasshouse, probably in Bohemia.

A fifteenth-century miniature of a forest glasshouse, probably in Bohemia, shows a slightly different plan, with a smaller annealing furnace attached to the main furnace (*fig 5*). Examples of this type of furnace, with a connecting annealing chamber, have been found as far afield as Sweden and the Baltic.[3]

Charleston points out that the rectangularity of the northern style furnaces was by no means the rule. The late Saxon site at Glastonbury in England appears to have been oval in plan, and oval ground plans as well as rectangular ones have been found on medieval glassmaking sites in Czechoslovakia.[4]

At some point, probably during the sixteenth century, the northern tradition diverged into two types: furnaces with subsidiary wing furnaces attached, and two-chamber furnaces where the working furnace and fritting/annealing furnace were part of the same structure.

Southern glassmakers appear to have favoured a round furnace with compartments ranged one above the other. The foundations of a circular furnace have been excavated in a seventh/eighth-century glasshouse on the Venetian island of Torcello,[5] and also at a late fourteenth/early fifteenth-century site at Monte Lecco in the Apennines.[6] The first detailed description of a southern furnace is given in a Syrian manuscript, dated not earlier than the ninth century AD, which is held by the British Museum. The three-storey furnace, with a lower fire chamber, a central melting chamber, and an upper vaulted annealing chamber which the manuscript describes, can be seen in the earliest picture of a glass furnace taken from another manuscript, *De Universo* (1023), the work of Rabanus Maurus, held in the library of Monte Cassino. These three-tiered furnaces were not always circular in plan, but could be rectangular, as the excavators of an eleventh/twelfth-century glass furnace at Corinth (Agora, South Centre site) discovered.[7]

A description which undoubtedly reflects the structure of contemporary Venetian furnaces is given in Vannoccio Biringuccio's *De la Pirotechnia*, first published in Venice in 1540. He describes beehive-shaped furnaces which comprise three storeys. The bottom two storeys consist of the firebox and the melting or founding furnace, and the top storey is the annealing chamber. He adds:

> Above this vault another vault is made which seals up and covers the whole; this is two 'braccia' [*ca* 110 cms] high above the first so that it completes the reverberatory furnace. This is the cooling chamber for the works when they have been made, for if they did not receive a certain tempering of air in this, all the vessels would break as soon as they were finished when they felt the cold . . .

Fig 6 A glass furnace with annealing furnace attached, from Johann Kunckel's *Ars Vitraria Experimentalis* (1679). A = vertical supports to furnace structure; B = 'tease-hole' (stoke hole); C = 'glory holes'; D = crucibles; G = conjoined annealing furnace; H = clay tunnels to contain glass for annealing.

In his book *De re Metallica* (Basel, 1556), the German writer Georgius Agricola described two differing furnaces. In the northern or 'German' type, known as the German furnace, the founding furnace was connected horizontally with a second chamber:

> At the back of the [main] furnace is a square opening, in height and breadth one palm, through which the heat may penetrate into the third furnace adjoining. This is oblong, eight feet [2.5 metres] by six [2 metres]

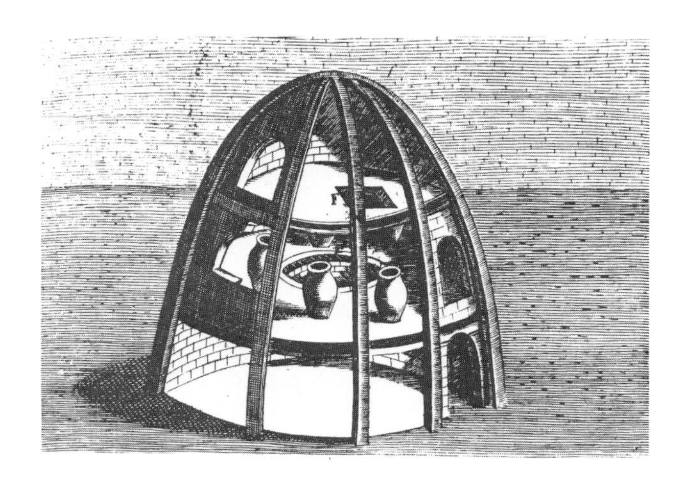

Fig 7 A Venetian-style furnace in section and complete, from Johann Kunckel's *Ars Vitraria Experimentalis.* A = lower fire chamber; B = 'glory hole'; C = annealing chamber; D = access hole for annealing chamber; F = hole allowing heat to enter annealing chamber.

broad, similarly consisting of two chambers, of which the lower has an opening in front for stoking the hearth. On either side of the stoke hole in the wall is a chamber for an oblong pottery tunnel . . . about four feet [1.25 metres] long, two feet [60 centimetres] high, and one and a half feet [45 centimetres] broad. The higher compartment should have two openings, one on either side, high and broad enough to admit the tunnels . . . in which the glass articles now made may be placed to cool off in a milder heat . . .

The second furnace described by Agricola is the so-called Venetian type – circular and beehive-shaped, with three storeys, the bottom compartment being used for burning the fuel, the middle compartment, where the heat was greatest, being used for the actual glass melting, and the top compartment, furthest from the heat, being used for annealing.

The two distinct types of furnace are again described in *Ars Vitraria Experimentalis*, published in 1679 by Johann Kunckel, the celebrated German chemist who ran the Potsdam glassworks. *Fig 6* shows the

Fig 8 A glasshouse, with a tiled shelter, in operation. From Johann Kunckel's *Ars Vitraria Experimentalis*. A = 'Tease-hole' (stoke hole); B = 'teaser' (stoker); C = main furnace; D = working platform; E = subsidiary furnaces.

northern type of furnace with annealing furnace attached, and *fig* 7 shows the Venetian type in section and complete.

Kunckel also included an illustration of a glasshouse complete with tiled roof and openings to let out the smoke from the furnaces, and in the background independent subsidiary furnaces can be seen with arched openings (*fig 8*). It is interesting to note the Continental practice shown here of working on a platform close to the furnace.

Between 1575 and 1662 the glassmaker's chair was introduced, and its distinctive shape, with long wooden arms, has remained essentially unchanged to the present day.

Raw materials

In northern Europe, which was heavily forested, glassmakers obtained their soda from the ashes of bracken, oak, beech and pine, as well as kelp. Their silica came from sands found in Lorraine, Silesia, and northern Bohemia – and from rocks: basalt, granite, obsidian and felspar are all known to have been used in Bohemia.

In the south soda was derived from the ashes of seaweed or other marine vegetation. From the fifteenth century onwards the glassmakers of the Middle East and southern Europe began to use the ashes of the glasswort (*salicornia herbacea*): its soda content is high, and it grows in abundance on the Mediterranean coast. It was often known by its Spanish name, '*barilla*'. As Venetian glassmaking techniques spread through the rest of Europe *barilla* became the favourite source of soda.

Silica was often obtained in the south by crushing pebbles from river beds. The Venetians used quartz-like pebbles from the Ticino and the Sile for their finest glass. The Swedish priest, Peder Månsson, who visited Rome between 1508 and 1524, described Italian glassmaking ingredients in his *Glaskonst* (Art of Glass).

> In Rome and Welshland [Italy] glass is made of three sorts of materials: fine white sand, black ashes, made by burning a plant which is there called kali or alkali and in Italian soda, and of a salt which is called sal alkali, the ashes of which are imported from Spain and from Alexandria and France to Rome for glassmaking, and likewise from other countries. The soda plant only grows on the seashore.

Cullet, or broken glass, was usually added to the silica and soda to speed up the melting process, since glass of all types melts at a lower temperature than its separate constituents. Although glassmakers used up their own glass waste in this way, they never produced enough for their own furnaces, and cullet was therefore bought and sold on a large scale.

THE RISE OF VENICE

As well as advancement in art and architecture, the Renaissance brought the expansion and development in Europe of luxury industries like silk-weaving, lacemaking, faience-production and metal casting, but the glass industry was apparently the first to be developed along new and fashionable lines.[8] Apart from stained glass windows, the products of medieval European glassmakers had not been outstanding. The decorative skills of enamelling, gilding, painting, engraving and tooling were virtually unknown to them, unless they had been fortunate enough to see the work of their Near Eastern contemporaries. Suddenly this static, conservative scene was transformed by the rediscovery of classical knowledge, combined with a dynamic new attitude to glassmaking. The focal point of this regeneration was fifteenth-century Venice.

Glass had been made in Venice itself since the tenth century; but as the recent excavation on the island of Torcello in the Venetian lagoon has shown (see p 63), the glass industry has been there even earlier, in the seventh or eighth century.[9] Venetian glassmakers had formed their own guild by 1268. Many of its members came originally from other northern Italian glassmaking towns – Treviso, Vicenza, Padua, Mantua, Ferrara, Ravenna, Ancona and Bologna.

By a decree of the Grand Council of Venice in 1292, glassmakers were encouraged to set up their businesses on Murano, a lagoon island about five kilometres north of Venice. This was a safety precaution, partly against fire, and possibly also against 'industrial espionage'. By 1300 the move was complete. The city took a grim view of any glassmaker attempting to leave and practise his art elsewhere; he was automatically condemned to death as a traitor.

Techniques

The Venetian and other Italian glassmakers of the Renaissance period re-introduced many old glassmaking techniques which had been lost to the West for centuries. Their rediscovery of old techniques was soon superseded by the invention of many new, exciting methods of making and decorating glass. Their main achievement was the invention of a fine, clear, colourless glass that they called '*cristallo*', but this was closely followed by the invention of many beautifully coloured glasses, and highly skilled decorative techniques, notably 'filigree' work.

Coloured glass

The Venetian glassmakers produced gloriously rich colours – blue, opaque white, green, purple and turquoise. Many formulae for coloured glasses are given in Antonio Neri's *L'Arte Vetraria*, including the recipe for a *red* glass, but it was not until the end of the seventeenth century that a clear ruby or pink glass could be made consistently well.

Plate 16 Large standing cup and cover in Venetian *cristallo,* enamelled and gilded. Second half of the fifteenth century. *Height 421 mm (16.5 in)*

The Venetians rescued *opaque white* glass from obscurity and used it to imitate much-coveted Chinese porcelain. It is probable that the first attempt to produce porcelain in the West, which was said to have been made by Maestro Antonio di S. Simone in Venice in about 1470, was an experiment with white glass. Opaque white glass was being made in Venice in the manner of porcelain by 1490, becoming popular in the early years of the sixteenth century, and was referred to as '*lattimo*'. Tin oxide was probably used to produce the white opacity in this early Venetian glass. Besides imitating porcelain, *lattimo* glass was also used in glass canes for filigree ('*vetro a filigrana*') decorated ware.

Glass with a semi-opaque or opalescent milk-and-water appearance, now called '*opaline*', was in production in Venice in the seventeenth century. Neri referred to Venetian opal glass as 'peach coloured', and tinges of blue, green or pink can be detected in some opaline glasses.

Continuing the ancient tradition of using glass as a substitute material for semi-precious stones, Venetian glassmakers invented an elaborate technique for producing glasses with the appearance of veined stones which they called '*calcedonio*'. The mingled, more or less opaque, colours were made to resemble agate, onyx and chalcedony, and the glass is sometimes mis-called by the German name '*Schmelzglas*'.

Cristallo

Muranese craftsmen apparently perfected their technique for making *cristallo* between 1450 and 1460. Angelo Barovier (*ca* 1400–60) was master of a Murano glasshouse and an acclaimed glassmaker, and it would seem from a document dated 21 February 1457 that he and another Muranese glassmaker, Niccolo Mozetto, were instrumental in working out the formula for *cristallo*, since they were granted the right to conduct work on 'crystal glass' outside their normal work by the *Podestà* of Murano.[10]

The Italian glassmakers had rediscovered the process of decolorizing glass with manganese, a process which had been lost elsewhere in Europe. Their ingredients easily fused together in the furnace and the metal that was produced lent itself to elaborate manipulation with pincers and tongs. However it was rarely without a tinge of brown, yellow or even a blackish tone.

Decoration: by the glassmaker

Venetian and other Italian glassmakers were highly skilled in the decorative techniques they had inherited from the past – trailing, mould-blowing, blobs and prunts – but they re-introduced and invented many new and complex ways of embellishing their wares. The Venetians expanded the *trailing* technique into stem-work and open work that resulted in the delightful, fragile-looking dragon-stemmed and winged glasses that were so popular in Europe in the seventeenth and even the eighteenth centuries (*pl 17*). They indulged in further extravagance of the trailing technique when they created their glass ships or 'nefs' which were made with elaborate open trellis work, with zig-zags of different coloured glasses built upon each other in a style reminiscent of Near Eastern 'dromedary' flasks of the sixth to eighth centuries AD.

Combing and marvering-in applied glass threads to decorate glassware was another ancient technique adopted by Italian glassmakers from their Egyptian, Roman and Islamic predecessors. Threads of opaque white glass (*lattimo*) were laid on to the surface of a vessel and combed upwards or downwards with a pointed instrument to form feathered patterns, and then marvered into the surface of the glass, leaving a smooth finish. Ewers, wine glasses and bottles were among the vessels which received this decoration. Venetian glassmakers also used the technique of marvering small pieces of glass into the body of a glass vessel.

Plate 17 *Flügelglas* – winged glass – from the Netherlands, late seventeenth century. *Height 299 mm (11.75 in)*

Plate 18 Sixteenth-century Venetian plate in *vetro di trina. Diameter 225 mm (8.88 in)*

The Venetians developed the technique of incorporating threads of *lattimo* into the body of a vessel. Thin rods of opaque white glass were caught up and worked very gently into a gather of clear colourless glass which was then blown to the desired shape. Alternatively, rods of *lattimo* and clear colourless glass would be laid side by side on a tray and fused together in a kiln, before being caught up on a gather of clear colourless glass. The technique spread to other countries with the Italian glassmakers, each country producing its own variations in the '*façon de Venise*'. Very rarely, other colours besides opaque white were used for threading, such as yellow, purple and blue. A superb sophistication of this technique was the Venetian '*vetro di trina*', where two plates with opposing white radiating thread decoration were fused together to form one piece. Tiny air bubbles were caught between each intersection of the threads, giving a delicate and rich appearance (*pl. 18*).

Ice glass was a Venetian innovation: a bubble of hot, clear, colourless glass was plunged for a moment into water, then reheated, resulting in a crackled, frozen appearance from which the technique derives its name. The same effect could be produced by rolling a bubble of glass on a marver which had previously been covered with fragments of broken glass, which adhered to the bubble.

Filigree glass was perhaps the greatest decorative invention of Venetian glassmakers, and the technique has remained popular to the present day. It is believed to have been created during the early years of the sixteenth century and is referred to in Biringuccio's *De la Pirotechnia*. By reheating and blowing an assembly of rods containing opaque white twists caught up on a gather of clear colourless glass, the Venetians formed vessels having a delicate white filigree pattern within their thin walls. The technique spread quickly to other European countries and remained long in use.

The ancient mosaic technique was resurrected by the Venetians in the form of '*millefiori*' glass. Although *millefiori* closely resembled Roman mosaic glass, it was produced by an entirely different method. The sections of glass rod were made in the usual way but were embedded in a gather of clear colourless glass or clear pale blue glass which was then blown to its final shape.

'*Avventurin*' glass (commonly spelt 'aventurine' today) was a Venetian invention which dates from at least the seventeenth century. In appearance it is usually blue or *calcedonio* with a sparkle which suggests sprinkled gold dust probably caused by copper reduced to its metallic form. It was sold by Italian glassmakers to foreign factories in rods or large pieces.

Decoration: by the decorator

Pure decoration, added after the glassmaker had finished his structural work, was to achieve great importance and high standards during the period of Venetian supremacy. *Enamelling* on glass was a speciality of the Italian artists, a technique they developed during the fifteenth century. From around 1470 expensive and large glasses were being decorated with richly complex figure designs, and from the early sixteenth century heraldic motifs became widely popular. Although their enamelling appeared similar to that of the Islamic artists, it would seem that the Venetians independently re-invented enamelling on glass, possibly borrowing the ide` from Italian metal workers. By tradition, the invention of Venetian enamelling has been ascribed to Barovier, one of the inventors of *cristallo*; but the Venetian scholar,

Luigi Zecchin, has shown that there is no evidence for this. After the sixteenth century enamelling passed out of fashion in Venice, apart from wares made for export, since Venetians began to appreciate glass in its own right, and looked for quality and beauty of shape rather than added decoration.

The glassmakers of Barcelona, great rivals of the Venetians, developed their own distinctive style of enamelling during the late fifteenth and early sixteenth centuries. Their motifs were Near Eastern in feeling, with stylized trees, arabesque foliage and running animals depicted in light yellowish-green, white, lavender-blue and black enamels, although occasionally human figures in contemporary costume also appeared.

Contemporary descriptions explain how enamelling was carried out: the cakes or beads of enamel were pounded on marble or porphyry, and the resulting powder was washed and applied to the already annealed glass vessel. The glass was then carefully reheated so that the enamels fused successfully. By the end of the seventeenth century it seems that enamelled vessel glass was beginning to be fired in special 'muffle' kilns, rather than in the furnace itself.

Schwarzlot, or black enamelling, originated in the Netherlands and spread to Germany, where it was improved by Johann Schaper and his followers in Nuremberg in the third quarter of the seventeenth century. Later the technique spread to Bohemian and Silesian glass-houses.

Although it had gone out of fashion in Italy enamelling remained a firm favourite with the Germanic peoples until the second half of the eighteenth century. Particularly popular were enamelled Elector glasses (*pl 19*) and '*Reichsadlerhumpen*' – 'Imperial Eagle beakers' – decorated with the Imperial double-eagle. One of the prescribed tasks for glassmakers wishing to join the guild at Kreibitz in Bohemia in 1669 was to prepare with colours an Imperial Eagle, with all its members, in one and and a half days'.

Cold painted glass achieved some popularity in Italy around the middle of the sixteenth century, when enamelling had fallen out of favour. Elaborate pictorial subjects, often based on Raphael's work, were painted in unfired oil colours on to the glass, but were naturally very liable to damage.

Gilding, in conjunction with enamelling, was frequently used by Venetian glassmakers, the gilding having a peculiarly light, soft appearance. Occasionally, gold decoration was used on its own, resembling a golden mist through which designs and inscriptions could be seen. Haudicquer de Blancourt in his *De L'Art de la Verrerie*

Plate 19 Bohemian *Humpen* in pale straw-tinted glass, enamelled and gilded, with designs showing the Emperor and Seven Electors, dated 1625. *Height 298 mm (11.75 in)*

(Paris, 1697) gives two rather over-simplified descriptions of the process of gilding. The second, and more convincing, method he describes involved painting the surface of the glass with gum-water, applying gold leaf and washing the leaf over with a solution of borax, then sprinkling on glass, ground to a very fine powder. When this was fired, the final layer of glass was allowed to melt, then the vessel was taken out and cooled.

Venetian glass *engraving* was a new decorative technique as far as Renaissance Europe was concerned, especially in its introduction of diamond-point engraving, a method probably borrowed from metalwork. In 1549 a Venetian glassmaker, Vincenzo dal Gallo, was granted the privilege of making engravings with diamond or flint stone on glass, which he had been requesting from the authorities since 1534.[11] The thin and brittle Venetian *cristallo* was particularly suited to the technique of diamond-point work, the glass taking the impress with precision yet allowing the engraver much freedom of movement.

THE VENETIAN INFLUENCE ABROAD

By 1500 the demand for Venetian glass had reached almost passionate proportions in Europe, and it became essential to the table of any person of social pretensions. The obvious advantages of having Venetian glassmakers operating in one's own country, rather than importing the costly wares, were quickly appreciated in northern European circles. Consequently, outside agents were used to entice Italian glassmakers to leave for other parts of Europe, and in spite of the heavy penalties imposed, particularly by Venice, glassmakers left the city in their hundreds between 1550 and 1700 to serve other European markets.

The most important rival to Venice was the industry conducted at Altare near Genoa. Originally started by glassmakers from Normandy, its tradition was similar to that of Venice, and, like that city, it rose to greater importance in the second half of the fifteenth century. Unlike the Venetians, the Altarists sought to spread the craft of glassmaking, and their craftsmen travelled far and wide throughout Europe, producing the fine glassware known as '*façon d'Altare*' or '*façon des Sieurs Altaristes*'. The glassware of runaway Venetians and Altarists and other Italian glassmakers from the north is practically indistinguishable one from another, and in many cases it is impossible to be sure whether a given vessel dating from the sixteenth and seventeenth centuries was made in Italy or by Italian glassmakers working in England, France, Germany, Spain or the Netherlands.

As early as the 1530s, Venetian and Altarist glassmakers are recorded as working in a glasshouse at Hall-in-the-Tirol, Austria. The important '*Hofglashütte*' at Innsbruck founded by the Archduke Ferdinand II (1520–95) was staffed by Venetians, and other Germanic rulers found it diverting to establish their own Italian glasshouses. In France there are ample records which prove the presence of Venetian and Altarist glassmakers from the late fifteenth century onwards.

The establishment of Italian glassmaking on a commercial basis notably took place in the Netherlands, which eventually became a clearing house for Italian glassmakers who went from there to staff other glasshouses in many parts of north-western Europe. Venetian-style glassmaking probably first started in the Netherlands at the beginning of the sixteenth century and was certainly well established by the middle. Italian glassmakers were at work in Britain as early as 1549, but it was not until fifty years later that Venetian glassmaking became firmly established here, the impulse coming from the Netherlands. Italian glassmakers were also operating in Copenhagen and Stockholm in the latter part of the sixteenth century.

In Spain, the Venetian styles were for a time superimposed on a native tradition which went back to Roman times, but towards the end of the fifteenth century clear colourless glass, commonly yellowish in tone, was being made to rival Venetian *cristallo*. Venetian styles were followed in Catalonia, especially in Barcelona, which had an extensive sea-borne trade with Venice and the Levant. The Venetian style was also followed at Cadalso near Toledo.

Glassmakers working in other countries under the influence of Venice often adapted and improved on Venetian techniques to suit local tastes. For instance, during the second half of the sixteenth century Nevers and Orléans became important centres of Italian-style glassmaking in France, Nevers sometimes being referred to as 'a little Murano'. The technique of building up figurines of coloured glass on a metal core, which originated in Nevers and spread to many other parts of France, was generally known as '*verre de Nevers*'. The French also adopted as their own the Venetian technique of marvering small pieces of glass into the body of a vessel: the resulting French 'marbled' or 'pebbled' glass was usually produced in bright opaque colours splashed on a light blue ground.

Although the technique of diamond-point engraving was practised in Italy, it was possibly more popular in other European countries, notably Holland and Austria, where it was usually used in conjunction with gilding. Motifs included fantastic birds and long-tailed monsters, coiled foliage and coats of arms. In the last third of the sixteenth century most European countries were producing glasses with this form of decoration with scrolled arabesque foliage, borders of chain or guilloche pattern, hatched 'ladder borders' and borders of single formal leaves or of cresting. In seventeeth-century Holland, diamond-point engraving became a fashionable pastime amongst amateurs, some of whom became very skilled at it.

The technique of applying *blobs and prunts* – hot drops of glass – to glass vessels as decoration is thought to have orginated in Syria. It was adopted by Venetian glassmakers when the Syrian wares were imported into Europe in the fourteenth century, but gradually became a speciality of Germanic glassmakers. The earliest known prunted

'*Stangenglas*' (literally 'pole glass') is said to have been found in the old quarter of Prague, and to date from the first half of the fifteenth century. These tall prunted vessels, known also as '*Spechtergläser*' and '*Passgläser*', soon became immensely popular, and remained so until the Baroque period (*pl 20*). The Bohemian pastor, Mathesius, remarked in a text of 1562 that the prunts were applied so that the glasses could be held easily, even by the clumsiest of people.

Plate 20 Early sixteenth-century German covered beaker in clear green glass with prunts. *Height 248 mm (9.75 in)*

By the middle of the fifteenth century the dropped-on spots of glass became larger blobs, broadly melted on the vessel and drawn out to a point. These vessels are variously called *Nuppenbecher* (drop beakers) *Warzenbecher* (warty beakers) or *Krautstrunk* (cabbage stem) glasses. The *Daumenglas* ('thumb glass') was further development of the Germanic blob technique: hot blobs of glass were applied to the surface of the heated vessel, and then the glassmaker sucked through his blowing iron so that the blobs were extended into the interior of the vessel, forming finger grips.

With virtually no opposition or competition from other countries, Venetian, Altarist and other Italian glassmakers managed to dominate the European trade in luxury glassware up to 1700. Their revival of old glassmaking skills and invention of many new techniques gave them a command of glassmaking which was to continue more or less unrivalled until the nineteenth century, despite the eclipse of their famous *cristallo* towards the end of the seventeenth century.

1. *Polak 1975*, 14–15
2. See *Charleston 1978* for details of furnace development
3. *Seitz 1939*, 279–82; *Roosma 1966*
4. *Charleston 1978,* 21–3
5. *Gasparetto 1965*; *Tabaczyńska 1970*
6. *Mannoni 1972*
7. *Charleston 1978*, 11–12
8. *Polak 1975*, 64
9. *Tabaczyńska 1970*
10. *Polak 1975*, 65
11. *ibid*, 54

CHAPTER FOUR

Glassmaking on the Continent since 1700

The seventeenth century was marked by a race amongst European glassmakers to find a new clear colourless glass, thicker and more robust than the Italian product and suitable for deep, rich engraving. Thick clear glass was increasingly in demand by the scientists of the seventeenth century for use in lenses. Cut decoration for hardstones was a major feature of the time, and it became essential to develop glasses that could stand the same deep cutting and engraving.

Almost simultaneously, during the 1670s and 1680s, Germany and Britain produced glasses which surpassed Venetian *cristallo* in quality (see pp 87, 118ff). Once the new products were on the market, incised decoration for vessels became the rule, whether deep cut, wheel-engraved, diamond-point engraved or stipple engraved.

From the end of the seventeenth century the new heavy cut glasses produced in Bohemia began to take over Venetian markets in Europe and Italian glassmaking went into a steady and inexorable decline, matched by the decreased political influence and importance of Venice. A brief revival of the craft occurred in Venice in the middle years of the eighteenth century, but by 1773 the glassmaking population of Murano had sunk to one tenth of its former strength – 383 glassmakers – and in 1806 the glassmakers' guild, along with all the other Venetian guilds, was formally dissolved.[1]

Little new glassmaking literature was published in the eighteenth century. Neri, Merrett and Kunckel remained unchallenged until well after 1800, and as late as 1797 the *Encyclopaedia Britannica* based its glass section on Neri's work. The fourth reprint of Neri's *L'Arte Vetraria* was issued in Milan in 1817. Christopher Merrett's translation and commentary on Neri's book, *The Art of Glass* (1662), was itself translated into Spanish in 1776. Kunckel's *Vollständige Glasmacherkunst* or *Ars Vitraria Experimentalis* (1679) contains German translations of Neri's book with Merrett's comments on it, together with Kunckel's own commentaries and additions. It had its third edition in 1743 and fourth in 1756, both issued from Nuremberg; and a French translation of the work appeared in 1752, published in Paris.

The steady and sometimes spectacular improvements in glassmaking techniques during the eighteenth century gave way to what has been called 'the golden age of glassmaking' during the nineteenth century. At no time since the Italian Renaissance had there been such an explosion in new glassmaking methods and ideas, and all over Europe there was keen rivalry in inventing new types of art glass for a highly competitive market. The day of the glass technologist had dawned, with large firms hiring scientists specifically to discover new methods of making and decorating glass. The demand for the new

Fig 9 The interior of a '*verrerie en bois*', illustrated in Diderot and D'Alembert's encyclopaedia (1772).

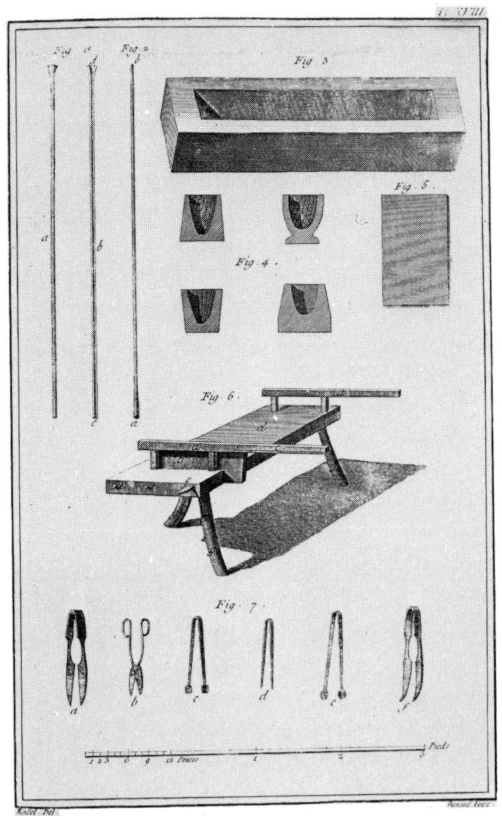

Fig 10 Glassmaking tools from a 'verrerie en bois', from Diderot and D'Alembert's encyclopaedia. 1 = blowing irons; 2 = pontil; 3, 4 = trough and moulds for shaping glass bubble; 5 = cast iron marver; 6 = glassmaker's chair; 7 = glassmaker's tools, including scissors, tongs, shears and pucellas.

glasses, which reached a far greater public than ever before due to cheaper production methods, reached its zenith at the turn of the century.

Fuels and furnaces

Wood-fired furnaces continued to be used on the continent until well into the nineteenth century, although there were occasional experiments with coal. French glassmakers tried using coal at Rouen in 1616 and at Namur in 1640; but it was not until 1829 that the 'Manufacture

Royale des Glaces' at St Gobain began to use a coal furnace for melting, and even then the 'firing' of the glass (the removal of bubbles of gas) was completed in a wood-fired furnace.

Diderot's encyclopaedia, published in France in the mid-eighteenth century, contains illustrations of both coal- and wood-fired furnaces. The high timbered building covering the furnace area of the '*verrerie en bois*' is shown in the interior view (*fig 9*). The fuel was burnt in the trough between the siege banks which supported the pots, and the products of the combustion passed directly over the pots, a process known as direct firing. In the same picture, a boy feeding wood into the fuel trough can be seen in the foreground, two glass-makers' chairs stand to left and right of the furnace with the 'gaffers' sitting on them doing the most complicated part of the shaping of vessels, and other members of the glassmaking team can be seen gathering, blowing and marvering glass around the furnace area. A long annealing arch is attached to the main furnace along which the glass vessels were moved on iron trays. Various glassmaking tools are depicted (*fig 10*), including shears, tongs and blowing irons, which are very similar to those used today. *Fig 11* shows a plan of a plate glass furnace, also taken from Diderot's encyclopaedia.

The coal-fired glasshouse was used in France to make bottle glass. Coal needed more oxygen for successful combustion, so increasingly elaborate flue systems evolved beneath the furnace to supply the fuel with the right amount of air. In the section of the bottle glasshouse which was fired by coal (*fig 12*), the deep flue can be seen below the central furnace with the grill upon which the coal burned. The glass-works is housed in a timber-framed building similar to the structures which covered the wood-fired furnace. The furnace itself has the traditional look of the wood-fired furnace, with platforms (sieges) for six glassmaking pots (crucibles), and four 'wing' or auxiliary furnaces attached.

All early furnaces had been direct-fired – the heat passed directly over the pots – but the nineteenth and twentieth centuries brought increasingly complex heating methods. Semi-direct-fired furnaces were introduced with the Boetius furnace, invented around 1865–70 and used in Europe and Britain well into this century. The method of preheating used in the Boetius furnace involved utilizing some of the heat from combustion, thus saving fuel.

The major break-through in furnace design came with the regenerative furnace or 'Wanne' invented and developed by Friedrich and Karl Wilhelm Siemens from 1856, which was applied to glassmaking by their brother Hans in Dresden by 1860, and was in general use

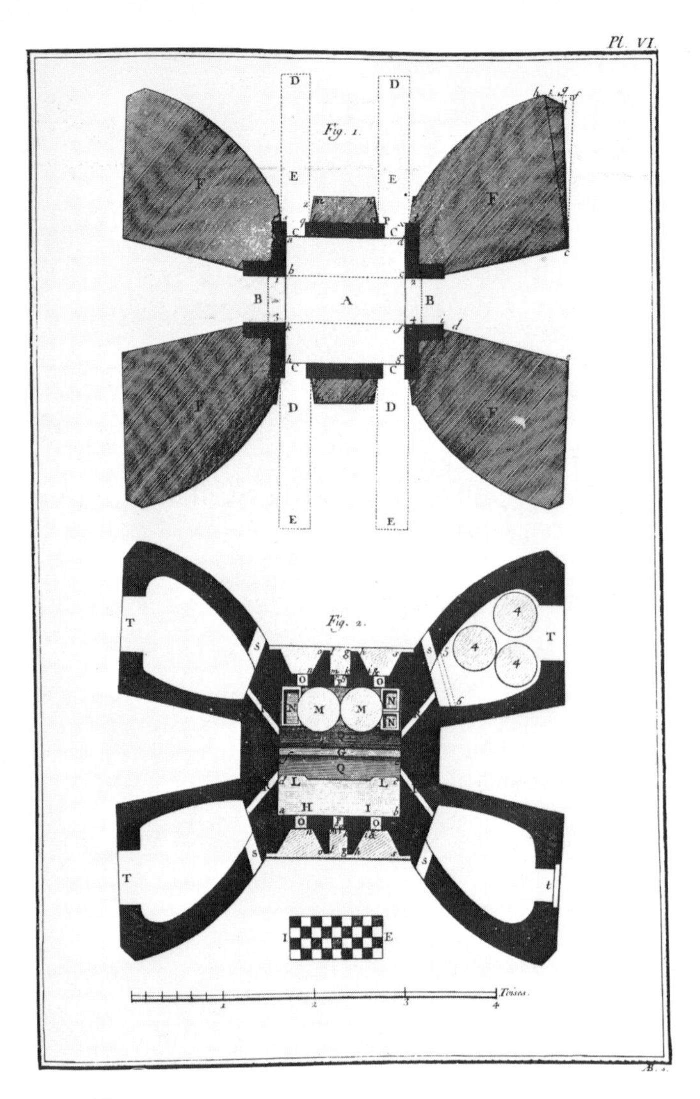

Fig 11 Plan of a plate glass furnace for four pots, with four 'wing' furnaces attached to the main one, from Diderot and D'Alembert's encyclopaedia. A = fire chamber; B = 'tease-holes' (stoke holes); F = wing furnaces; G = space between sieges; HI = siege for crucibles; L = siege for 'cuvettes' (rectangular clay containers into which the glass was ladled before removal from the furnace); M = crucibles; N = cuvettes; O = openings through which glass was transferred from pots to cuvettes; T = pot arches (for firing crucibles); t = fritting furnace.

Fig 12 Section through a coal-fired bottle glasshouse, from Diderot and D'Alembert's encyclopaedia. a = furnace roof; b = space between sieges; c = grille for coal fuel; d = crucibles; e, f, g = working outlets; i = subsidiary furnaces.

throughout Europe from the 1870s. The earliest 'reverberatory' furnace reflected the heat from the fuel down from the curved dome of the furnace on to the pots, thereby increasing the temperature; in a regenerative furnace hot gases were passed through four chambers, known as regenerators, beneath the crucibles, the passage of the gases being regularly reversed. This process resulted in very little heat loss, a constant temperature, and a considerable saving in fuel costs.

Regenerative tank furnaces, in which crucibles were replaced by one large tank of glass, continuously flowing with hot metal to the glassmaker or his machines, were used to make flat glass and bottles. They effectively introduced the cheap, mass-produced glassware that is familiar to us today. Regenerative pot furnaces were also used to produce finer products, such as tableware.

Friedrich Siemens' original design used solid fuel burnt in fireplaces built into the furnace, but the smoke and dust managed to permeate the glass through the regenerators, causing contamination of the glass batch. This problem was solved by introducing gas-firing: gas was made in a separate unit known as a 'producer' and fed into the furnace.

The regenerative furnace has remained in use to the present day, with only relatively minor changes made to the original designs.

Raw materials

The flourishing glassmaking industry of northern Europe continued to use potash obtained from the ashes of local wood as a flux; but after 1861 it began to be mined at Stassfurt in eastern Germany, making western Europe independent of other sources.

In 1787 Nicholas Leblanc perfected the Leblanc process for heating sea salt to produce what was known as 'black ash', or sodium carbonate, which was then used by glassmakers for over fifty years. Salt cake, produced in the first stage of the Leblanc process, where common salt and sulphuric acid were heated together to produce sodium sulphate, was popular with glassmakers in the 1830s and 1840s as a cheaper form of alkali, although it demanded higher melting temperatures. The introduction of Solvay soda in Belgium in 1863, where an ammonia-soda process produced sodium carbonate, re-established soda as the major source of alkali for glassmaking.[2]

From the nineteenth century, glassmakers tended to become more selective in the sands they used, so that staining from impurities became less likely, and in many cases was almost completely eliminated (see pp 111ff). The early nineteenth century saw the recognition of lime as an important stabilizing ingredient, largely as a result of the increasing use of pure manufactured alkali. The potash produced from marine and woodland plants had contained lime, alumina and magnesia without any need for the introduction of extra chemicals.

Clear colourless glass

One of the main aims of the Elector of Brandenburg, Friedrich Wilhelm (1620–88), and his protégé, Johann Friedrich Kunckel (*ca* 1630–1703), was to make at their glass factory in Potsdam a glass that rivalled Italian *cristallo*. It seems Kunckel succeeded in this at a very early stage, for he mentions 'how one makes a truly perfect crystal' in his famous book *Ars Vitraria Experimentalis* (1679), written only one year after his arrival in Potsdam. Unfortunately, Kunckel does not divulge his formula or method for this glass 'as we Germans now make it in various places', but stated his willingness to inform special friends.

This solid clear glass developed in Germany and Bohemia in the 1670s was made by adding lime to stabilize the purified potash glass. The potash replaced the traditional soda, and a large proportion of chalk provided the lime ingredient. The resultant glass was indeed crystal-like, thick and strong enough to take the deepest engraving.

The French *Compagnie des Glaces,* set up in 1665 in Paris, was brought to a standstill when its staff of eighteen Italian glassmakers refused to do the work in the manner required. After this, several glasshouses staffed only with Frenchmen were set up in various parts of the country to make flat glass, suitable for windows and mirrors. By 1670 they were capable of producing a fine 'white' glass, and the import of foreign mirrors was forbidden two years later.

In Sweden, although the Kungsholm glasshouse in Stockholm was founded in 1676 for the production of glass in the Italian style, by the turn of the century the factory was making glass in the Bohemian-German manner – clear colourless glass with engraved decoration – and continued to do so until the factory closed in 1815. Engraved glass was also produced at Henrikstorp between 1691 and 1760, and the famous factory of Kosta in Småland was founded in 1742. A few glasshouses were also set up in other parts of Sweden during the second half of the eighteenth century.

Finland's first glasshouse was set up in 1681 in Nystad, but larger and longer-lasting establishments were founded in the next century at Åvik (1748), Mariedal (1779), Thorsnäs (1781), Nyby (1783) and Notsjö (1793).[3] Norway's Nöstetangen glassworks produced clear colourless glass *à la façon de l'Angleterre* under the influence of an English glassmaker, James Keith, who worked there from 1755 to 1787.

Other English and German glassmakers emigrated further afield, to North America, and set up manufactories on the north-eastern seaboard. Little is known of the earliest ventures, during the seventeenth century, but more successful glasshouses were founded in the eighteenth century, notably by the German Caspar Wistar, at Co. Salem near Philadelphia, in 1739–80, by Henry William Stiegel, originally from Cologne, at Elizabeth (1763) and Manheim (1765–74), Pennsylvania, and by John Frederick Amelung at New Bremen, Maryland in 1784.

Clear colourless glass was far less important in the nineteenth century than previously. The countries of Europe continued to produce excellent clear glass with engraving and cutting of increasing complexity – new ventures in first-class crystal glass included that of France's René Lalique (1860–1945) – but Europe was outdone by North America during its so-called 'brilliant' period of clear colourless cut glass (*ca* 1880–1915).

Coloured glass

In spite of the popularity of the cut and engraved clear colourless glasses produced in the latter part of the seventeenth century in Europe, coloured glasses were still in demand. A consistently good *red* glass was only developed at the end of the seventeenth century. Deep red could be made with iron mixed with a little calcined brass, but the finest reds were produced by using copper in the mix.

Gold ruby glass was not entirely the invention of Kunckel, though he certainly considered it one of his major achievements during the time he directed the glasshouse at Potsdam. He was too wary to commit his recipe to writing, but nevertheless '*Goldrubinglas*' was soon being produced by Bohemian and German glassmakers. An eighteenth-century Potsdam formula describes what may have been the Kunckel method. A gold ducat was beaten thin, cut into pieces and dissolved in a heated mixture of one and a half ounces of spirits of salt (hydrochloric acid), half an ounce of nitric acid and one dram of sal ammoniac. This solution was added to the glass mix which was melted in the usual way. The vessels made from the glass were subsequently reheated to develop the colloidal dispersion of gold particles within the glass which produces the red colour.

Though Kunckel is given most of the credit for producing this much more reliable gold ruby glass, others were experimenting in the same field and may possibly have come up with the same results at approximately the same time. Some time before 1679, Andreas Cassius of Leyden had found that if he added tin chloride to gold chloride solution a purple powder was precipitated which, when added to the glass mix, could produce a red glass after reheating.[4]

The vogue for Chinese, and later European, porcelain in the seventeenth and eighteenth centuries increased the demand for *opaque white* glass painted in the same way. France and Germany, and possibly even Spain (though there is a likelihood that much opaque white glass found in Spain was imported from Central Europe) produced opaque white glass in imitation of porcelain. The same styles and even the same artists were apparently sometimes used to decorate both porcelain and opaque white glass products, the most popular motifs being Chinese designs, birds and flowers.

Green glass of two distinct types was produced in Europe over this period. Besides the expensive and purposely coloured rich green glass of the Italian glassmakers, the more 'common' or everyday glass continued to be made in large quantities for the expanding bottle industry, coloured in a huge variety of greenish and brownish tints by iron impurities in the mix.

During the *Biedermeierzeit* in the Germanic countries, the period that followed the Napoleonic wars, coloured glass gained a new popularity. The central figure of this movement away from the traditional cut, clear colourless glass of the preceding centuries was Friedrich Egermann (1777–1864), an outstanding manufacturer of coloured glass at his glasshouse in Blottendorf, near Haida in northern Bohemia. Patronized by Count Kinsky, he launched his famous 'Lithyalin' glass in 1828, which continued the ancient technique of *imitating the appearance of semi-precious stones* with its marble-like aspect. He also introduced other coloured glasses, notably '*Chamäleon-Glas*' in 1835, which was imitated in both Bohemia and Silesia. *Opaque turquoise* glass was produced in many parts of Bohemia during this period; and *black* 'Hyalith' glass was produced in Glatzen, in southern Bohemia, at glasshouses belonging to George Franz August Longueval, Count von Buquoy.

'Cased' glass, where two or even three gathers of different coloured glasses were taken up on the blowing iron and blown and shaped as one, became increasingly popular in the nineteenth century. Lighter stains or flashings (an almost skin-like covering compared to the heavier casing) of ruby-red and yellow glass were often used on clear colourless glass in the same way.

Yellow glass was produced in the same manner as in medieval times, using a silver compound which gave a yellow stain to the surface. New shades of greenish-yellow and yellowish-green with an opalescent sheen were made with uranium, first by Josef Riedel of Isergebirge in the 1830s, including colours named 'Annagelb' and 'Annagrün' after his wife. George Bontemps of the glass factory of Choisy-le-Roi was producing similar glass from uranium in about 1838. Eugène Rousseau (1827–91), another French designer, produced coloured glass which resembled precious stones – a nineteenth-century revival of an ancient tradition.

Decoration

The new heavy glasses developed in the Germanic countries in the late seventeenth century were unsuited to the decorative techniques used

Plate 21 Covered goblet in clear colourless glass with *Hochschnitt* engraving, probably Silesian, early eighteenth century. *Height 321 mm (12.63 in)*

on the delicate and rather brittle Italian *cristallo*. At first, Italian styles were continued with the new glasses, but before long new techniques were developed to complement the strength, heaviness, thickness and clarity of the Bohemian crystals. Contemporary tastes turned away from the 'adding' decorative techniques of trailing, enamelling and gilding to the 'taking away' techniques: cutting, wheel-engraving, diamond-point engraving and stipple engraving. Engraved designs began to cover practically every part of a glass vessel and new, more powerful engraving machinery had to be developed.

Cutting and engraving are basically the same technique: a rotating abrasive wheel cuts into the glass surface. For deep cutting, large wheels are used, and the worker holds the glass on top of the wheel, between himself and the tool. For engraving, small wheels are used, and the engraver holds the object below the wheel so that he can see what he is doing. Cutting is the term used for large-scale geometric designs, usually relatively deeply incised, and engraving is used for fine, detailed and usually pictorial work.

The new potash-lime glass developed in Bohemia and Germany in the late seventeenth century was eminently suited for the technique of wheel-engraving. During its most flourishing period (*ca* 1685–1775), German glass engraving was done mainly by unknown independent craftsmen working in north-eastern Bohemia and Silesia. The best work was done by the engravers to three German courts, notably by Friedrich Winter (*d.* before 1712) for Count Christof Leopold Schaffgotsch at Kynast in the Riesengebirge, not far from Hirschberg in Silesia. Brother of Martin Winter, the Court Engraver at Potsdam, Friedrich's work was much admired by the Count, who installed a cutting-mill for him at Hermsdorf, below the Petersdorf paper mill. It was powered at first by horse, but in 1690–1 it was converted to water, which was more effective for *Hochschnitt* (deep cutting) work.[5]

Martin Winter (*d*. 1702) set up an engraving workshop in Potsdam near Berlin in 1687 which was also run by a water-powered mill. Towards the end of the seventeenth century, Landgrave Carl of Hesse set up up a water-powered cutting mill at Kassel where both hardstones and glass were engraved in relief by Franz Gondelach (*b*. 1663), sometimes called the greatest German master of the art, who became '*fürstliche Glasschneider*' to Carl in 1695, and was still described as '*Hofglasschneider*' at Kassel in 1716. Both *Hochschnitt* and *Tiefschnitt* (intaglio work) were used to decorate the German glass thickly, and relief cutting was the sole ornament used on some of the more spectacular glasses produced in Silesia at the end of the seventeenth century (*pl 21*).

Steam power to drive cutting mills appears to have been introduced generally in Europe towards the end of the eighteenth century. A picture of a steam-driven cutting-shop appears in Dionysius Lardner's *Cabinet Cyclopaedia* of 1832, where '. . . steam power is used for giving motion to a shaft which causes the revolution of numerous large wheels or drums fixed thereon, and each of these being connected by a band with a pulley on the axle of a smaller wheel, occasions the latter to revolve with great celerity: these small wheels are the cutting instruments' (*fig 13*). Water and fine sand were fed on to the wheels to provide an abrasive, and wooden wheels, 'the edge dressed with either pumice stone or rotten stone' and finally putty powder, were used to impart the final polish.

The Bohemian workshops remained supreme in the late seveneenth and early eighteenth centuries, but after *ca* 1725 Silesia overtook them in importance. Some of the best Silesian work was done by a Warmbrunn artist, Christian Gottfried Schneider (1710–73). By the beginning of the eighteenth century wheel-engraving on glass was becoming increasingly popular in Holland, its best-known exponent being Jacob Sang, a Saxon who worked in Amsterdam producing signed works between 1752 and 1769.

A new method of glass decoration developed in the Netherlands was *stipple-engraving*, in which grouped and graded dots were engraved with a diamond point on the surface of a glass object, the dots representing the highlights of the design. The diamond point was set in a handle which may have been gently struck with a small hammer to produce individual dots on the glass. Frans Greenwood (1680–1761), a native of Rotterdam who held an official position in Dordrecht from 1726, produced a quantity of stippled glasses, often signed and dated.

Aert Schouman, G. H. Hoolart and J. van den Blijk also practised the art of stippling, but the most famous name connected with this technique in the last forty years of the eighteenth century was that of David Wolff (1732–98), who was born at 's-Hertogenbosch, married at the Hague in 1762 and continued to live there until his death.

Decoration in the nineteenth century

Glass decoration was stimulated in the nineteenth century not only by the new technological discoveries, but also by the revival of the Venetian glass industry from the 1830s onwards. By 1832 the first filigree glass was being produced by the descendants of some of the old Muranese glassmaking families. In 1846 '*calcedonio*' was again being made in Venice. By the middle of the century, *avventurin* and *millefiori* had been rediscovered and in 1864 the first exhibition of '*vetraria Muranese*' was put on show in the new glass museum in Palazzo Giustinian. The large international exhibitions which were such a feature of that century always included Muranese glass in bright colours, with maximum use made of gold dust. By the mid-nineteenth century most other European glass factories were again imitating Venetian techniques of glass manufacture.[6]

The Venetian style, unlike the traditional cut glass that had been fashionable elsewhere, made much more of the inherent ductile and manipulative qualities of the material glass. This must have inspired John Ruskin when he wrote about glass in the appendix to the second volume of his *Stones of Venice* (London, 1851–3):

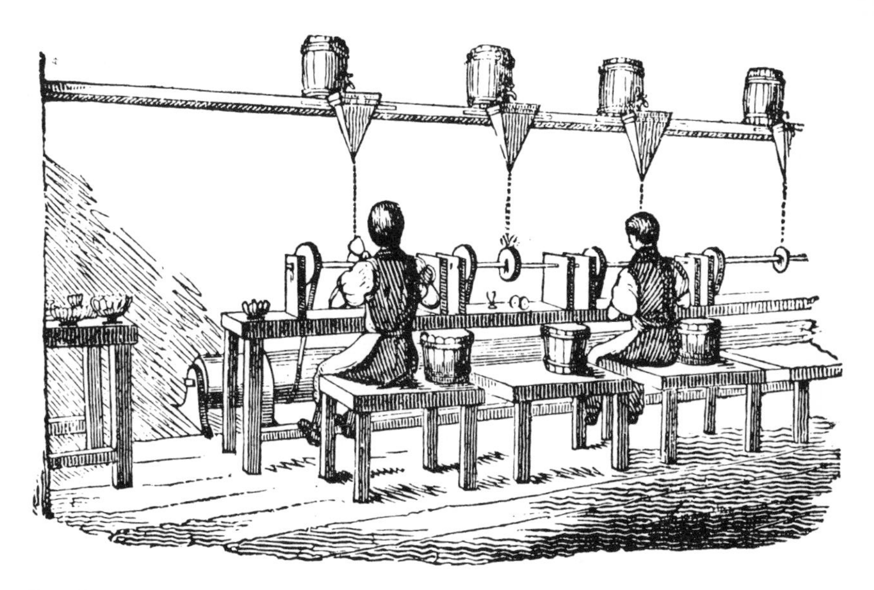

Fig 13 Drawing of a cutting shop, from Dionysius Lardner's *Cabinet Cyclopaedia* (1832)

> In its employment for vessels, we ought always to exhibit its ductility, and in its employment for windows its transparency. All work in glass is bad which does not, with loud voice, proclaim one or other of these qualities. Consequently, all cut glass is barbarous, for the cutting conceals its ductility and confuses it with crystal. Also, very neat, finished and perfect forms in glass are barbarous.

Large-scale commercial production of glass in traditional Venetian styles started in 1866 when Antonio Salviati (1816–90) set up his own furnace in the Palazzo Mula. Many other Muranese glassmakers followed him in making pastiches of sixteenth and seventeenth-century Venetian glass, aimed predominantly at the tourist market. At the same time there was a limited production of simpler glass articles in Venice. After the First World War, Functionalist ideas gave a new stimulus to the traditionalism of Venetian glassmaking and a truly modern style was established, notably by Paolo Venini and Ercole Barovier.

The nineteenth century saw increased sophistication in the art of *enamelling* glass. Samuel Mohn (*b*. 1762 in Weissenfels, *d*. 1815 in Dresden) and his son, Gottlob Samuel Mohn (*b*. 1789 in Weissenfels, *d*. 1825 in Vienna), working under the patronage of the Imperial family circle, decorated glasses with translucent enamelling. Their chief technical innovation was the preparation of transparent enamels, in contrast with the heavy opaque enamels used particularly in the sixteenth and seventeenth centuries. The Mohns' discovery of transparent enamels was further improved by the Viennese porcelain and glass decorator, Anton Kothgasser (1769–1851) who produced some of the finest examples of the Viennese *Biedermeier* style.

Gilding in the form of gold sandwich glass – '*Zwischengoldgläser*' – was reintroduced in Bohemia in the eighteenth century, the best specimens dating around the 1730s, though less skilled work continued until about 1755 (*pl 22*). It was still being made at the end of the eighteenth century by Johann Joseph Mildner (1763–1808) of Gutenbrunn in Lower Austria. The technique known as '*verre eglomisé*', where gold or silver leaf was fixed to the back of a sheet of colourless glass and etched with a point, goes back to the fourteenth century. It

Plate 22 Clear colourless glass beaker made by the *Zwischengoldglas* technique, Bohemia, 1730s. *Height 79 mm (3.13 in)*

was used towards the latter end of the eighteenth century and into the nineteenth century, notably by an Amsterdam artist, Zeuner.

Lampworking, where glass objects were made by melting multi-coloured glass canes at a little lamp heated by a tallow flame, was notably used by the German etcher and painter, Karl Köpping (1848–1914), who designed some tall decanters and goblets made at the lamp from the tube glass between 1892 and 1900. His work formed part of the German *Jugendstil* movement.

Glass engraving and cutting was practised with as much skill in the nineteenth century as in the eighteenth. The *Biedermeierzeit* brought a new interest in glass engraving, characterized by vessels with massive feet and bold polygonal facetting, and very fine engraving on the better pieces. Glass engravers tended to move about and work independently, notably Dominik Bimann (1800–57) who went from Prague to Franzensbad during the season, and August Böhm (1812–90) who travelled to England and America to practise his art.

Cameo glass was produced in Venice in the nineteenth century using the Italian soda-lime metal. French glassmakers produced their own variation on cameo work, using acid etching to engrave their designs on to blanks of cased coloured glass. Their designs were strongly influenced by the Orient. The most famous name associated with French cameo glass is that of Emile Gallé (1846–1904), who established his own workshop for glass decoration in 1867 and, with his father, began the regular production of art glass in Nancy in 1874. The firm of Daum in Nancy, established in 1875, produced fine cameo glass with flower and tree designs rather than figures. There were many other lesser known workers in French cameo glass and their methods and designs were copied throughout Europe.

Diamond-point engraving on glass was practised in the nineteenth century by Andries Melort of Holland (1779–1849) who copied the work of Dutch painters on to flat sheets of glass. D. H. de Castro (*d.* 1863) who was a chemist in Amsterdam, revived the technique in the middle of the century.

Acid etched glasses are known to exist from the seventeenth century, but the process was not generally used in glassmaking until the nineteenth century, with the discovery of hydrofluoric acid. An acid resist such as wax paraffin would be put over the surface of the glass and the desired design cut through the resist to expose it to the acid. The glass would then be plunged into an acid bath. The technique could be used to cut through one layer of glass to another in the case of cased, flashed or stained glass, or to give a single layered glass a matt finish.

As the name implies, *sand-blasting* entails directing a stream of sand, crushed flint or powdered iron on to the surface of a glass in a jet of air. A stencil plate of steel may be used to protect the areas to be left plain, or an elastic varnish or a rubber solution painted on to the glass. The type of finish can be varied from matt to very rough by altering the size of the nozzle, the abrasive or the air pressure. The technique has been in use since 1870, almost invariably to decorate flat glass articles such as windows and mirrors, though occasionally it has been used for lettering on mass-produced items of vessel glass.

The technique of *pâte de verre*, which was known in antiquity and was re-introduced into the glassmaking repertoire by the Frenchman Henri Cros (1840–1907), is a curious cross between pottery and glassmaking, whereby a plastic material of powdered glass can be made into sculptural forms or vessels by a process of moulding. After many years of experiments, Cros finally succeeded in producing his own *pâte de verre* and made a famous series of reliefs in this material between 1893 and 1903. His son Jean followed in his father's footsteps, but Henri Cros never disclosed to anyone else the actual constituents he used. Albert Dammouse (1848–1926), Georges Despret (1862–1952), Emile Gallé (1846–1904) and François Décorchement all found their own individual interpretations of *pâte de verre*.

The nineteenth-century development of *mechanical pressing*, which received its greatest impetus in the U.S.A., revolutionized the glass industry and allowed the humblest person to possess glass objects. Even the physical appearance of glass changed: it emerged from the pressure-operated moulds with a dull surface, which manufacturers sought to hide by covering it all over with decoration. Expensive cut glass was an obvious target for cheap pressed glass imitations, but new combinations of motifs also appeared, culminating in the twenty-five years of American glassmaking known as the 'Lacy Period' because of the emphasis on all-covering detailed decoration (1825–50). In the 1870s and 1880s Victorian English glass took on some of the characteristics of the U.S.A.'s Lacy glass.

The old technique of blowing glass into moulds composed of several sections was improved in America, where iron moulds were introduced. As well as being durable, these could be carved with the most intricate patterns. Mould seams or slight ridges running round a glass vessel reveal whether a two, three or four-piece mould was used.

Twentieth-century glass

The twentieth century has seen a flowering of talent in European glassmaking. Among the many people who have produced outstanding work are France's Michel Daum of Nancy, who brought a new style of almost abstract vessel shapes to the famous Daum factory after 1945; Maurice Marinot, who was perhaps the most influential personality in the history of art glass between the wars, and directed the production at the factory of Viard, near Troyes; and René Lalique of Wingen-sur-Moder who became famous for his colourless, often matt-finished glass.

Bohemian designers have been internationally acclaimed, especially Professor René Roubíček, who has been chief supervising artist at the glassworks Borské sklo; Jaroslav Brychtová; Stanislav Libenský; Pavel Hlava; and Jan Kotik. Although Czechoslovakia is best known for its cut and engraved work, skilful furnace-worked glass has been produced by people like Emanuel Beránek at the Škrdlovice factory and František Zemek at the Plame König factory.

Sweden's great contribution to twentieth-century glassmaking is epitomized in the imaginative glass produced in the Smålands glassmaking area, where the Strömbergshyttan, Åfors, Kosta, Skruv, Boda and Orrefors factories flourish. In particular, the designs of Simon Gate and Edward Hald of Orrefors have been internationally acclaimed. Finland's Gunnel Nyman established a free-flowing style of glass which was taken up and developed successfully by Finnish designers. The simple designs of Timo Sarpaneva are among the most outstanding.

The work of Wilhelm Wagenfeld and Wilhelm v. Eiff of Germany has influenced the products of the whole industry. Chris Lebeau was the pioneer in Dutch art glass, ably followed by Adrian Dirk Copier, Floris Meydam and Willem Heesen. Among the most important glass artists in recent times have been the Italians Paolo Venini and Flavio Poli and the Austrian Stephen Rath.

Modern European glass designers tend to use the natural ductile and manipulative qualities of glass to achieve their effects, rather than added decoration, although the old cutting and engraving techniques remain popular.

1. *Polak 1975*, 68
2. *Douglas and Frank 1972*, 19
3. *Polak 1975*, 161–62
4. *Douglas and Frank 1972*, 19
5. *Charleston 1965*, 42–44
6. *Polak 1975*, 202

CHAPTER FIVE

Glassmaking in Britain

Some authorities would trace British glassmaking back to the Bronze Age, on the evidence of faience beads that have been found on graves and at other sites in for example Wiltshire and Wessex. But in fact it is impossible to tell whether these were made locally or imported from the Mediterranean.

Iron Age sites have yielded other glass objects besides beads: a number of glass armlets and a set of playing pieces were found for instance at Welwyn Garden City; but again there is nothing to indicate that these were not foreign imports. Bronze objects made during the Iron Age, such as shields, helmets, armlets and harness mounts, are occasionally decorated with red glass inlays, sometimes mistakenly called enamels. These were almost certainly applied from a lump of glass which was softened and pressed into the inlay cavity.[1] The red glass was probably imported from abroad in the form of cakes or slabs.[2]

It has generally been assumed that the glass used in Britain during the Roman period was imported; and indeed the major finds at Sittingbourne, Kent (*pl 23*), at Wint Hill in Somerset, and at Barnwell in Cambridgeshire seem to have come from a variety of sources including Syria, Egypt, Italy, Gaul and the Rhineland. However, a few Romano-British glassmaking sites have now been found, indicating that at least utilitarian and window glass were being produced in this country. The first of these was the Wilderspool glasshouse in Warrington, Cheshire. Since its discovery, Roman glasshouse sites have been found at Mancetter, Warwickshire, Caistor St Edmund in Norfolk and Middlewich in Cheshire, and Roman glasshouse debris was found at Wroxeter (Viroconium), Salop in 1972.[3]

From the period following the Roman withdrawal, in the early fifth century, we have only a few isolated pieces of evidence of British glassmaking. Glass bracelets as well as *millefiori* glass and beads have been found in Britain and Ireland, dating from this Early Christian period. A millefiori glass rod found at Luce Sands, Wigtownshire, suggests that similar glassmaking may have been going on in south-west Scotland as well. A unique type of triangular yellow bead with black stripes has been found in Pictland. Although somewhat circumstantial, the evidence from Dinas Powis, Glamorgan, implies glass working there in the seventh century AD.[4] Harden has mentioned the possibility of a pagan glass-working in Kent[5] and a late Saxon glass-working site has been found at Glastonbury.[6]

Despite the lack of evidence of glassmaking during this period a surprising number of glass objects has turned up on British Dark Age

Plate 23 Large mould-blown flagon with strap handle in clear green-tinted glass, found at Sittingbourne, Kent. Roman Empire, late first century AD. *Height 311 mm (12.25 in)*

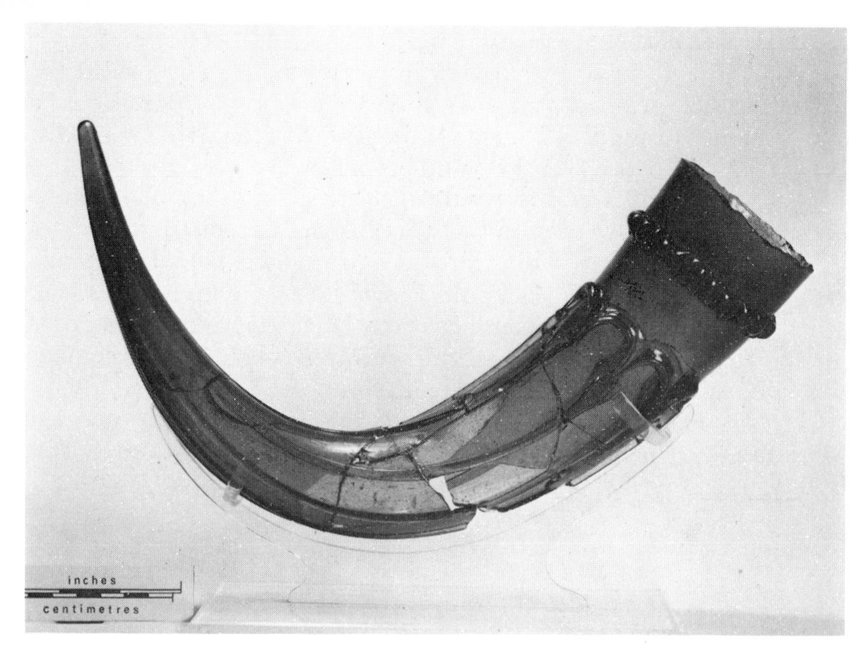

Plate 24 Drinking horn found in 1937 on the site of a Saxon cemetery at Rainham, Essex. Late fifth or sixth century AD. *Length across from tip to rim 330 mm (12.9 in)*

sites, many probably of Belgian or northern French origin. While paganism continued (fifth to seventh centuries), glasses were still deposited in graves, and Roman-style vessels such as horns, palm cups and cone beakers continued to be manufactured (*pl 24*). Looking rather like a green glass ice cream cone, and decorated with exquisite trailing, the British Museum's cone beaker from Kempston, Bedfordshire is one of about twenty that have been discovered on British sites (*pl 25*). Claw beakers, probably produced in the Seine-Rhine area in the late fifth and sixth centuries, have also been found on British sites, notably at Castle Eden, Taplow and Mucking. Blue glass trailed vessels, including squat jars, found at Broomfield in Essex and Cuddesdon, Oxfordshire, were probably produced in England (*pl 26*). Other local products included bell- and bag-shaped beakers, pouch bottles and bowls, possibly made in Kent (*pl 27*).

One of the first acts of Archbishop Theodorus, who was enthroned on 27 May 669, was to restore Wilfrid to the Archbishopric of York. 'Wilfrid immediately proceeded to act with characteristic munificence. He found his cathedral dilapidated, and he restored it. The thatched roof he covered with lead; the windows, hitherto open to the weather, he filled with glass; and such glass . . . as permitted the sun to shine through'.[7] Wilfrid, Bishop of Worcester, is reputed to have glazed his church at the same period.

Where the two Wilfrids got their glaziers from the documentary sources do not say; but in 676 Benedict Biscop, founder of the Abbey at Monkwearmouth, Northumbria, had to send to Gaul for glassmakers to make windows. According to the Venerable Bede's *Opera Historica*, written in the seventh century, 'When the work was drawing to completion he sent messengers to fetch makers of glass, who at this time were unknown in Britain, to glaze the windows of the church, its side chapels and clerestory'. The glassmakers from Gaul 'not only finished the work required, but taught the English nation their handicraft, which was well adapted for enclosing the lanterns of the church, and for the vessels required for various uses'.

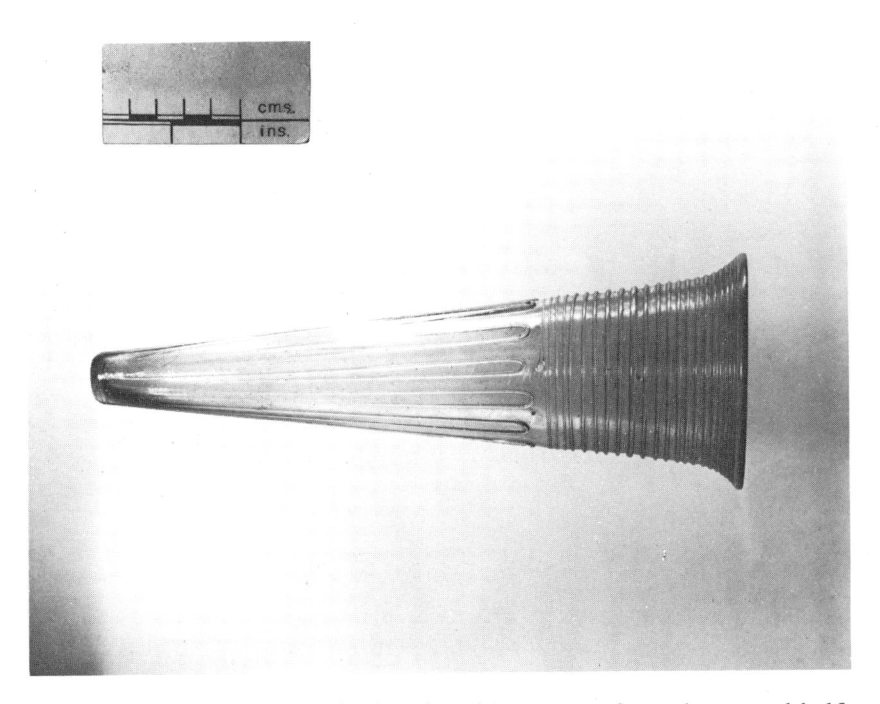

Plate 25 Green glass cone beaker found in a grave from the second half of the fifth century AD at Kempston, Bedfordshire. *Height 262 mm (10.3 in)*

Plate 26 Seventh-century AD dark blue squat glass jar with zigzag trailing from a rich grave at Broomfield, Essex. *Height 73 mm (2.8 in)*

Evidently the attempt to teach the British to make glass was a failure, for in 758 Cuthbert, Abbot of nearby Jarrow, sent an urgent appeal for glassmakers to Archbishop Lullus of Mainz in the Rhineland. 'If there be any man in your diocese who can make vessels of glass well, pray send him to me; or if by any chance he is beyond your bounds, in the power of some other person outside your diocese, I beg your fraternity that you will persuade him to come to us, for we are ignorant and helpless in that art; and if it should happen that any one of the glassmakers through your diligence is permitted, D.V., to come to us, I will, while my life lasts, entertain him with benign kindness.'

Archaeological evidence of this Northumbrian glass industry at Monkwearmouth and Jarrow was found in the late 1960s by Rosemary Cramp. Both sites produced window glass, plain and coloured, of a generally similar type, though with minor variations. The glass, which was soda-lime with a comparatively large iron content, was bubbly and markedly thin when compared to Roman products. Both sites revealed pale blue, olive green, amber, yellow-brown, red and green-blue glass. Monkwearmouth also produced emerald green and

Plate 27 Trailed pouch bottle found near Bungay in Suffolk. Seventh century AD. *Height 123 mm (4.8 in)*

dark cobalt blue colours. Among the remains, which included crucible fragments, Cramp also discovered irrefutable evidence that *millefiori* glass was produced at Jarrow.[8]

We have documentary evidence dating from the twelfth and thirteenth centuries which has been said to prove the existence of a glassmaking industry in other parts of Britain. The title 'Vitrarius' or names like 'Verir', 'Verrer', etc, may indeed refer to makers of glass; but they may alternatively mean 'glazier'. The earliest mention comes in a medieval manuscript *The Story of Bacton*, from Bromholm Priory, Norfolk. It tells us that Henry Daniel, a 'vitrarius', was made Prior of the influential monastery of St Benet's at Holme, Norfolk, by King Stephen (1135–54). A deed which cannot date from later than 1240 grants twenty acres of land (probably at Dyers Cross in Surrey) to one 'Laurence Vitrearius'. Another deed, relating to lands nearby, was granted to 'William le Verir', thought to be the son of

Laurence Vitrearius, in 1300. At about the same time the records of jurors in Colchester reveal a family called 'le Verrer'.

It is not until 1351 that we find solid documentary evidence of British glassmaking. In the Exchequer K. R. rolls, between 30 April and 12 December 1351, there are four entries recording the purchase of 'white' glass from John Alemayne of Chiddingfold, Sussex, for St George's Chapel, Windsor, and St Stephen's Westminster. In 1355, 480 pounds of glass was purchased for Windsor from John Alemayne; and in 1378, John Brampton, third Master of the Glaziers Guild from 1373–81, was paid thirty shillings with carriage for a seam (120 pounds) of white glass of the Weald for the windows of the King's chamber. John Alemayne's property at Hazelbridge Hill, just south west of Chiddingfold church, was leased to John Schurterre, a 'glasier' in 1367. Chiddingfold seems to have remained the centre of Weald glassmaking during the rest of the fourteenth century.

When John Schurterre died, in about 1379, his widow, Joan, brought John Glasewryth from Staffordshire to help run the family business, which indicates that there was a glass industry of some sort already in operation in Staffordshire at this date. Her son, John, was described as a 'glasier', and he had a glasshouse in Kirdford, a parish adjoining Chiddingfold, in 1385.

Although the domestic glassmaking industry was now established, luxury glassware was still being imported from abroad. In 1399 Richard II issued letters patent to Andrea Zane and Jacopo Dandolo, masters of two Venetian galleys then moored in the port of London, allowing passengers to sell their 'small wares' on deck, including glass vessels. Fragments of Italian-style glassware dating from the late thirteenth century have been excavated recently in Boston, Lincolnshire and Southampton and Italian-style glass from the late fourteenth to early fifteenth century has also been found, notably at Old Sarum, Wiltshire and Knaresborough Castle, North Yorkshire. Clearly British glass was still considered by many to be inferior to foreign products. In her will of 1439 Isabel Countess of Warwick specified that John Prudde of Westminster, who was painting the windows of her husband's chapel, should employ 'no glass of England, but glass from beyond the sea'. And in 1449 Henry VI invited John Utynam of Flanders to come and make coloured glass for Eton College and King's College, Cambridge, granting him a monopoly for twenty years. In about 1420, however, John Glasman of Rugeley bought white glass for York Minster which had been made in Staffordshire. And Salisbury Cathedral is recorded as having had its own 'glashous' in the late fifteenth century.

The glassmaking industry of the Sussex Weald presumably continued throughout the fifteenth century; but we have no documentary record of it between 1400 and 1495, when a certain Thomas Shorter conveyed property to Henry Ropley 'glasscaryour'. There is then another gap in the records until 1536, when John Petowe, a member of another well-established Chiddingfold family, left a will bequeathing his son John '10s of suche things as shall come and be made of the glasshowse and all my toyles (tools) and moulds as belongeth to the glasshowse.' It is interesting that both the Schurterre deed of 1380 and the Peytowe will of 1536 mention items dealing with vessel glassmaking, confirming that both window and vessel glass was produced on these early Weald sites.[9]

Domestic window glass became far more common in the sixteenth century than previously. Whereas before glass windows had been installed almost exclusively in churches and the houses of the rich, they now became available to the lower classes. A recent study of medieval Bristol notes that the Rose tavern in the High Street had glass windows by 1504/5. A weatherboard was provided for a glass window in a tenement in Redcliffe Street, one of Bristol's larger leasehold properties, in 1534. Lavish footage of window glass, some purchased in London, was provided for a pair of new houses built by the Corporation in 1548. By 1566, an inventory of moveables in a fairly modest house in Redcliffe Street, which was chiefly concerned with panelling and glazing, shows that while domestic window glass was still rare enough to be counted as a moveable asset, the house had a glass window over the stairhead in the hall, one in the small chamber fronting the street, another in the chamber over the parlour, and three in the kitchen.[10]

Nevertheless it is clear from Charnock's *Breviary of Philosophy* (1557) that British glasshouses were still few and far between:

> As for glassmakers, they be scant in the land
> But one there is as I do understand
> And in Sussex is now his habitacion
> At Chiddingfold he works of his occupacion.
> To go to him it is necessary and meete
> Or send a servante that is discreete
> And desire him in most humble wise
> To blow thee a glass after thy devise.

Although raw materials, fuel and a potential market were readily available, the missing factor in British glassmaking in the sixteenth century was skill – a commodity which could only be obtained from abroad.

Foreign glassmakers in Britain

It was Henry VIII who first made glass fashionable among the British nobility, when in the early sixteenth century he purchased large quantities of Italian *cristallo* for his royal residences. Inventories of noble households before this time confirm that the British preferred silver and gold plate for their tables, which had the advantage of being virtually unbreakable and of being a portable kind of wealth. But a 1542 inventory of Henry VIII's effects lists as many as 371 glass vessels, some richly finished in gold and silver mounts – ample evidence that luxury glass had now developed from a novelty to a vogue. The local greenish forest glass was no longer good enough for the comfortable Tudor manor houses that had replaced the castles of the Middle Ages. Clear colourless glass, as well as luxury glass vessels, were in demand and the domestic industry could not cope.

The first attempts to bring foreign glassmakers to Britain were abortive. In 1549 Edward VI offered a contract to eight Murano glassmakers to make glass in England, providing them with the capital to build and equip a glasshouse. When the Venetian Council of Ten recalled them and they attempted to leave, the king found it necessary to lock the glassmakers in the Tower until they agreed to continue to work for eighteen months in England before returning to Murano. In 1552 a London merchant, Henry Smyth, secured a patent to make 'brode[window] glass such as is wont to be called Normandy glasse' for twenty years. The remains of a Normandy-style glasshouse dated around 1550 have been discovered in the vicinity of land once owned by 'Henry Smith' of Burgate, at Knightons, Alfold, Surrey, and it seems possible that this was the site of Smyth's glasshouse (see p 136). However, there is no further documentary evidence of the project.

In 1565, Cornelius de Lannoy obtained a three years' contract to introduce to England the art of glassmaking 'as practised in the Netherlands', as well as certain privileges for schemes including the turning of base metals into gold. His capabilities did not match up to expectations and it was reported to Lord Cecil after a few months that 'All our glasse makers cannot facyon him one glasse tho' he stoode by them to teach them . . . The potters cannot make him one pot to content him'. Lannoy put the blame for his failure on the lack of good potmaking clay.

The credit for establishing a successful Continental-style glassmaking industry in Britain is variously given to Jean Carré, a French merchant, and to Jacob Verzelini, a Venetian glassmaker. Jean Carré, a native of Arras who had lived in the Low Countries for some years, settled in London with his family in 1567. By July of that year, he had already secured a licence from Queen Elizabeth I to build furnaces for making window glass in the Weald, and another from the Mayor and Alderman of London to build a Venetian-style furnace there. Carré gained freedom from local competition in the making of window glass for twenty-one years, when he and his partner, Anthony Becku, were granted letters patent, dated 8 September, 1567.

With the help of his French business associate, Jean Chevallier, Carré went ahead with his plans to bring foreign glassmakers to England to man his new furnaces. A contract was signed in April 1568 with Thomas and Balthazar de Hennezell of Lorraine to come to England with four other 'gentlemen glassmakers', to work in Alfold, a village on the Surrey-Sussex border where Carré had sited two of his furnaces. Peter and John Bungar from Normandy arrived the same year and worked, making crown window glass, at the second Alfold furnace. However when Becku insisted that the foreign workmen should train English apprentices the Hennezells departed angrily, and the fiery Bungars attacked two of their employer's relations. An appeal to the Privy Council by Becku in the same year resulted in an official investigation, where it was reluctantly concluded that there was no hope of Englishmen being trained by the foreign glassmen. Becku gave up any more attempts to engage actively in the industry, but gained some monetary compensation.[11]

The first Italian glassmaker known to have been connected with Carré was Ognia ben Luteri, who was in London in 1571. He later worked for Jacob Verzelini and is known to have had a furnace at Burgate in Godalming parish, near Guildford.

With the death of Carré in 1572, the fortunes of the small industry he had established took a downward turn, although it continued to be run for at least another four years by his in-laws and his business associate, Peter Briet. The high costs of bringing foreign workmen to England, building new furnaces, and overland transportation, had taken their toll on the business, and the arrogant attitude of the immigrant glassmakers helped to destroy local goodwill.

Jacob Verzelini (1522–1606) was a Muranese glassmaker who had worked in Antwerp for twenty years before he came to London in about 1565. He started running a *cristallo* glasshouse at Crutched Friars, an old monastery building near the Tower of London, soon after Carré's death. By 1574 he owned the furnace in London: but, as an alien, his position was insecure, so on 15 December of that year he applied for, and was granted, a monopoly. This granted him the sole right to produce drinking vessels in the Venetian style and metal for twenty-one years, and no foreign imports were to be allowed (*pl 28*).

The London glass-sellers resented the monopoly, and it may have been their action that caused a disastrous fire at Crutched Friars in September, 1575, which burnt the glasshouse to the stone walls. Nevertheless, Verzelini survived the blow, took the precaution of becoming an English citizen in 1576, which entitled him to own property, and rebuilt the glasshouse. The new furnace prospered, and his patent was not successfully challenged until 1592. More Italian glassmakers joined Verzelini, as well as a glass-engraver, who may have been Anthony de Lysle. Verzelini retired in 1592 to Downe in Kent, where he owned much land. He died in 1606 at the great age of eighty-four years and was buried at the parish church at Downe, where until recently a brass tablet commemorated him. A

Plate 28 Diamond-engraved colourless glass goblet with gilded and opaque white decoration. Dated 1586 and attributed to the London glasshouse of Jacob Verzelini. *Height 130 mm (5.1 in)*

small number of glasses attributed to Verzelini's glasshouse, decorated with diamond engraving sometimes attributed to Anthony de Lysle, still exist.

The St Bartholomew massacre in Paris in 1572 and the massacre of hundreds of Protestants in Antwerp in 1576 caused a large-scale immigration of foreign skilled workers into Britain. Among these were gentlemen glassmakers from Lorraine, notably the families of Hennezell (Hensey), Tyttery (Tittery), Tysack and du Houx, who were soon making glass in the Weald and at Buckholt in Hampshire. A few more Norman and Flemish glassmakers joined the foreign influx, increasing the competition for the Bungars, who had left Alfold and set up in Wisborough Green by 1579. Norman glassmakers who started to make glass in England were the de Cacqueray (Cakery) and Le Vaillant who are recorded at Wisborough Green (1600) and Buckholt (1576) respectively.

A glasshouse at Rye was shipping glass to London by 1574, and another at Northiam, East Sussex, was doing the same between 1580 and 1584. A glasshouse near Hastings was in operation around 1580, and Buckholt, Hampshire, had at least one glasshouse between 1576 and 1579, manned by foreign glassmakers. Some of the glassmakers moved to Knole in Kent and Ewhurst in East Sussex a few years later.[12] The foreign glassmakers at first stayed in the south of the country, but later small groups of families spread west and north, settling notably in Staffordshire and eventually in Newcastle on Tyne. Whether the aggression of local inhabitants towards these foreigners who used up so much wood as fuel for their furnaces had much to do with their migrations about the country is difficult to say. Certainly there was no sudden mass exodus from the Weald; from time to time glassmakers returned there to work, and they appear to have moved frequently and freely about the country.

Staffordshire, particularly the Bishop's Wood area between Eccleshall and Market Drayton, attracted most of the foreign glassmakers who chose to leave the south before 1590. Stourbridge clay, which was most suitable for potmaking, was a strong attraction, but a more direct invitation to come to the area was probably given by the Bishop of Lichfield, who had been rector of Balcombe in Sussex until 1579, and no doubt appreciated the profits to be made in supplying fuel to the glass industry. Glassmaking at Bishop's Wood was carried out by the Henseys, Titterys and Tysacks for at least twenty years between 1584 and 1604, and across the border in Cheswardine, Shropshire, references to the Bigos and du Houx appear in the parish registers between 1601 and 1613. Ambrose and Edward Hensey were involved in setting up a glasshouse on Sir Richard Bagot's Staffordshire estates in 1585.

Although the main concentration of the forest glass industry would appear to have been in the Weald and Staffordshire, occasional historical and archaeological evidence indicates that the glassmakers were willing to go to any area, no matter how remote, as long as it could supply the materials necessary for their craft. As far away as Bickerstaffe in Lancashire in 1600, the parish registers record: 'A stranger slayne by one of the glassemen beinge A ffrenchman then working at Bycerstaff and buried 10 December, 1600'.

Evidence of forest glassworking has been found in Ruyton-XI-Towns, Salop; Biddulph, Staffordshire; and east of Kenilworth, Warwickshire. The remote North Yorkshire glasshouse sites of Rosedale and Hutton were producing good quality forest glass in the late sixteenth century. No doubt many more sites remain to be discovered.

Besides vessel and window glass, these forest glasshouses were producing other specialized items, including distilling apparatus, hanging lamps, so-called 'linen smoothers' and hourglasses. Among the vessels, beakers with milled edged bases, and moulded, trailed and prunted decoration, were common products. Coloured glass is comparatively rare, but an opaque red glass was produced at Lower Chaleshurst, Chiddingfold, and Malham Ashfold, Wisborough Green in the Weald.

Monopolies

Foreign glassmakers were in control of both luxury and forest glassmaking in Britain in the late sixteenth century, and probably would have remained so, had it not been for the intervention of Queen Elizabeth I. The granting of industrial monopolies, as this scheming monarch soon realized, was an ideal way of rewarding courtiers and earning their devotion without incurring any expense to herself. Although in about 1590 she summarily turned down the request of two of her footmen for the sole right to make 'urynalls, bottles, bowles, and cuppis to drink in' which Verzelini was not producing, in 1592 she granted a new patent to Sir Jerome Bowes, a soldier, courtier, and her ambassador to Russia – but certainly no glassmaker. He was given the right to make *cristallo* products for twelve years, but the hostility of Verzelini's sons caused his new glasshouse in part of the former Blackfriars monastery to run into difficulties. Two city men, William Turner and William Robson, used their finances to help Sir

Jerome, and in 1598 Francis Verzelini was put in prison for ten years, where he was soon joined by his brother, Jacob, and financially ruined.

By 1601, the Blackfriars glasshouse was flourishing, probably employing most of Verzelini's former staff. With the departure of William Turner to the alum industry in Yorkshire in 1605, control of the monopoly lay effectively in William Robson's hands. This aggressive entrepreneur successfully fought off any threats to his (although nominally Sir Jerome Bowes') monopoly in the luxury glass trade, bought off the one real threat that came along in the form of Edward Salter's *cristallo* glasshouse, built at Winchester House in Southwark in 1608, and made sufficient profit from his Blackfriars glasshouse to be able to pay both Bowes and Salter a handsome rent for the privilege. He remained in a strong position in the luxury glass trade until 1612, when the Crown's support for the changeover from wood fuel to coal for glass furnaces ended his supremacy.

A tight grasp over the Wealden glass industry, the practical equivalent of a monopoly, was engineered by Isaac Bungar, 'the truculent Bungar', son of Peter Bungar whom Carré had brought from Normandy. He managed to gain control of the London flat glass market with the aid of Lionel Bennett, a glass dealer, and formed an alliance with his biggest rival in the Weald, Edward Hensey of North Chapel.[13]

FUEL AND RAW MATERIALS

The most important factor in the establishment of a new glasshouse was access to fuel; this took precedence over all other considerations, such as the availability of sand and alkalis, fireclay, and proximity to markets. Clearly the great advantage of the major British glassmaking areas, the Weald and Staffordshire, was their extensive woodlands. Beechwood and oak were particularly suitable for the glassmaker's needs.

Sand

The second most important factor was sand, which accounted for up to 75% by volume of the glass mix. As Christopher Merrett noted in 1662, different kinds of sand varied enormously: some would soon melt and mix with the other ingredients, but some would not. It was not, however, expensive: Merrett observed that 'it costs but little except for the bringing by water' to the London glasshouses. The Wealden glassmakers relied on local supplies of sand from the outcrops at Hambledon Common and Lodsworth Common. Washed sand from Lynn and sand from Alum Bay in the Isle of Wight were mentioned in an English glass recipe book of 1778–9. Sand from Lynn is still used for glassmaking today because of its great purity.

John Virgoe of Pilkington Research and Development Laboratories points out that glassmaking sands can be divided into two principal types.

a) Geologically recent deposits, known as 'drift', which occur as relatively thin superficial veneers in many valleys and low-lying areas, and some elevated areas. The Shirdley Hill sand used by Pilkington Brothers in the St Helens area is of this type, as also is the Chelford sand of Cheshire. Many of these deposits are mixed sands and gravels and are also used in building.

b) Solid, bedded sands and sandstones, usually of considerable thickness and forming the bedrock. These sands may be unconsolidated or hard, the latter needing crushing and grinding before use. The two principal sandstone formations used in the United Kingdom are the Millstone Grit, used by Pilkington Brothers from Oakamoor, Staffordshire, and the Lower Greensand worked for glassmaking sands in Surrey and at King's Lynn.

The classic publication on glassmaking sands is P. G. H. Boswell's *Memoir on British Resources of Sands and Rocks used in Glassmaking* (London, 1918).

Fluxing Agents

The extreme variation in the quality of English forest glass can be attributed to the inconsistencies of the various fluxes used, as much as to any other factor. Forest glassmakers kept the valuable ash left by the wood they had burned as fuel, using it as a fluxing agent in their next glass batch. Their habit of mixing together whatever ashes were available led to considerable variations in the glass they produced. Merrett commented that 'for green-glass in England they buy all sorts of ashes confused one with another, of persons who go up and down the country to most parts of England to buy them'. Besides the ash from beech and bracken, glassmakers tried other plants such as pea, bean and gorse with varying degrees of success.

Once coal furnaces were established in the early seventeenth century, other sources of alkali had to be found. A more refined flux was in any case necessary for the making of 'cristallo' *Barilla* (the ashes of the glasswort) was imported to England from Alicante, Spain, by Carré, Verzelini and Mansell in the late sixteenth and seventeenth centuries, Verzelini ordering 'Sevonus ashes' from Antwerp when his Spanish supplies were interrupted in 1589. Soda, barilla, saphora, rochetta, polverine and natron were all imported from the Mediterranean to England at this period to be used as fluxing agents. The State Papers for 1666 reveal that the Duke of Buckingham was using saltpetre (potassium nitrate) as a flux for crystal glass.

Potash was found to be more suitable for lead glass than soda, but first the impurities in crude potash had to be removed. By mixing the ashes in water and repeatedly boiling them, then evaporating the liquid in shallow bowls, a much purer pearl ash could be obtained, which may have been one of the alkalis used in fine English glass. American pearl ash was imported to England in the eighteenth century. Kelp from Scotland and Ireland, as well as England, was also used as a flux by British glassmakers.

Lime

The addition of fluxing agents to the silica could bring down the melting point of the mix to below 800° C, but such a low melting composition could easily be attacked by water. This could be remedied by the addition of lime, which made the glass more durable, and resistant to water attack. However an excess of lime could lead to crystallization. The ancient glassmakers had to gauge the correct proportions of ingredients most carefully to obtain a good glass.

Lime was seldom added to a mix deliberately before the end of the seventeenth century, since enough of it was introduced already via the major batch materials, or from corrosion of the crucible in the furnace. Merrett observed that when a larger proportion of purified alkali to sand was used, the glasses would fall to pieces if left to stand in damp places. He did not realize that the process of purifying the alkali had removed the essential lime and magnesia. It was not until the second half of the eighteenth century that the importance of lime as a stabilizing ingredient was realized, when P. Delauney Deslandes, Director of the St Gobain glass factory from 1758 to 1785, restored durability to glasses made from purified soda by progressively adding lime to a proportion not greater than 6%. Once pure manufactured alkali became generally available in the nineteenth century, lime took its place as a separate major constituent in the making of glass.

The term 'soda-lime' applied to so many ancient glasses rarely indicates the separate addition of lime by the early glassmakers. The high proportion of lime often revealed by analyses could be introduced in the sand and alkali. Vegetable ashes contain large amounts of lime, alumina and magnesia, all important stabilizing ingredients, and it was considered bad practice to add yet more lime which could lead to deterioration of the metal.

Refractories

A good fireclay was also essential. The crucibles and furnace structure needed to be able to withstand great temperatures, and a burst pot in the furnace could cause a great deal of damage from molten glass. There are few references to refractories in historical records, although Christopher Merrett, writing in 1662, did mention that pots for making 'cristallo' glass were made from Purbeck clay, and that the green glass manufacturers of London used clay from Nonsuch in Surrey, mixed with a clay brought from Worcestershire (the latter probably referring to Stourbridge clay, which later came into widespread use in Britain). Clay was often transported over long distances, and tests on crucible fragments from Wealden sites show they were made from good quality fireclay which was not local.[14]

The change to coal

The rate of fuel consumption for an Elizabethan forest glasshouse has been variously estimated at 700 to 1000 cords a year.[15,16] (A cord is a quantity of cut wood, approximately 128 cubic feet, which was originally measured with a cord or string.) Large tracts of woodland were completely stripped by landowners selling fuel to glassmakers, as at Petworth and Knole Park in the late sixteenth century. Between 7 June 1585 and 18 January 1586, 543 cords of wood were carried to the Knole Park glasshouse, which appears to have been in continuous production for a period of 226 days. The cost of wood varied throughout the country, but the heavy demand from glassmakers and iron makers was forcing the price up, encouraging some glassmakers to move to areas where fuel was less costly.

In 1589 one enterprising gentleman, George Longe, tried unsuccessfully to persuade Queen Elizabeth to suppress all English glasshouses and give him the monopoly for making glass in Ireland – where wood was fairly cheap – in order to save England's fuel. The London merchants of the Virginia Company also unsuccessfully attempted to establish a glasshouse at Jamestown, Virginia, in 1608–9 and 1621–4. By this time it was clear that the devastation of woodland could not be allowed to go on. Timber was vital for shipbuilding, as well as being used in the iron and glass industries. It was essential to find a new kind of fuel.

The first patent to experiment with the use of coal in many industries, including glassmaking, was issued in 1610 to Sir William Slingsby and others for twenty-one years. Sir William's success was not notable, and in 1611 a patent dealing exclusively with producing glass with coal fuel was given to Sir Edward Zouch, Bevis Thelwell, Thomas Mefflyn, the King's Glazier and Thomas Percival.

Robson vigorously opposed the Zouch enterprise, but King James I gave his support to Sir Edward because of the saving in wood. In 1614 he issued a new patent which granted Zouch the right to make every type of glass with coal fuel, revoked all previous patents, and forbade the use of wood as a fuel in glass furnaces. Robson made glass in Ireland between 1614 and 1618, but Zouch successfully enforced his rights against the importation of Robson's glass to England and Robson was forced to close down his Blackfriars glasshouse in 1614. Prosecution of provincial forest glassmakers commenced successfully in the same year. James I lent further weight to the prohibition by his proclamation of 23 May 1615. He noted 'there hath beene discovered and perfected a way and meanes to make Glasse with Sea-cole, Pit-cole, and other Fewell, without any manner of wood, and that in as good perfection for beauty and use, as formerly was made by wood'.

Understandably, the British glassmakers put up a strong fight to keep the fuel with which they were familiar, claiming that they only used the 'lop and top' of a tree, and left the main part undamaged. When they abandoned this feeble argument and tried to change over to the new fuel they found that coal had definite disadvantages over wood, for the heavier smoke and dirtier atmosphere coloured and spoilt the molten glass in the open crucibles, the fumes and smoke coming from the working holes of the furnace made working conditions difficult for the glassmen, and coal required more oxygen to burn properly, the draught in the wood furnaces being quite inadequate.

But once these technical problems had been overcome (see p 143ff), British glassmakers were in a position to exploit the superiority of their coal-fired furnaces, which could produce hotter temperatures and finer glass more consistently than ever before. The cone-shaped British glasshouse was to become a familiar sight during the late seventeenth and eighteenth centuries, leaving nations like the French still struggling into the early nineteenth century to master coal. The prohibition of wood-using may have appeared a harsh measure at the time, but it brought long-term benefits which have remained with the British industry until the modern period.

Sir Robert Mansell

Sir Robert Mansell, a Vice-Admiral in the English navy, who was to dominate the English glassmaking scene from 1615 to 1642, joined Zouch's company on the issue of its third patent early in 1615. The new patent was very similar to the previous one, but gave stronger emphasis to the banning of imports and the use of wood for fuel. During the summer of 1615 Mansell bought out his partners and thus gained complete monopoly of the British glass industry.

A new furnace for window glass had been built at Lambeth in 1613, which allowed Mansell to concentrate his Winchester House glasshouse on the production of *cristallo*, and to build a further *cristallo* furnace at Broad Street. The *cristallo* works, managed by William Robson and manned by new recruits from Italy and the Low Countries as well as indigenous workers, proved to be a financial success and produced mirrors and spectacle lenses of good quality, besides vessel glass. For window glass and cheaper green glasses, Mansell had to find a cheaper fuel than the Scottish coal he used in his London furnaces. He closed his Lambeth works down in 1615 and tried to find suitable locations at Kimmeridge in Dorset on the Isle of Purbeck, at Milford Haven, Dyfed and at Wollaton, near Nottingham but all failed for technical or financial reasons.

Bungar made the most of Mansell's inability to supply the flat glass market, and gained permission to continue producing glass at his wood-fired glasshouses until 1618, when his furnaces were finally suppressed. Mansell next established coal-burning furnaces at Newcastle-on-Tyne and new works at St Catherine's, London and Woolwich to supply the London market, in case his Newcastle works met with any difficulties. He sought to supply the rest of the market by a process of subleasing, where he allowed independent glassmakers to produce glass with coal, and had achieved reasonable success by 1620.

A new patent was issued to Mansell in 1623, further extending his monopoly of the industry. In 1630 the importation of drinking glasses and mirrors was forbidden, since most of Mansell's staff from Broad Street had fled across the Channel to help set up a rival *cristallo* establishment with Mansell's chief clerk, Vecon. At this time, Mansell's monopoly was supplying the English market adequately in flat glass, bottles, spectacles and common green glass vessels, and his depleted staff at Broad Street produced simple *cristallo* glasses and mirrors.

In 1635, Charles I granted Mansell an indenture prolonging his monopoly in England, and granting him a monopoly in Ireland for a rent of £1500. The high rent led to higher glass prices, and the quality of Mansell's glass deteriorated: but now for the first time England was exporting glass abroad and did not have to rely on imports.

In 1640 the Scots invaded Newcastle, which abruptly ended glassmaking there, and cut off the supply of Newcastle coal to Mansell's London furnaces. The House of Commons ordered Mansell to surrender his patent in May, 1642, but this did not deter him from rebuilding and operating his Newcastle works. Even when new competitors set up two new furnaces in 1645 it did not upset his business unduly, as his petition to the Common Council in 1652 for the extension of his leases demonstrates. He died soon afterwards and the Henseys and Tysacks eventually took over his furnaces.[17]

The Civil War and the Commonwealth

These were unsettled times for the British glass industry: luxury glassware was equated with the despised royalty, drinking glasses with drunkenness and mirrors with vanity. Nevertheless, the industry was lucrative enough for Cromwell to collect taxes from it.

George Villiers, second Duke of Buckingham (1628–87) took an interest in glassmaking amongst his other multifarious activities, and his name provided a useful umbrella under which less powerful glassmakers could operate. Following the arrival of Charles II at Dover in May 1660, an enterprising French official, John de la Cam, persuaded the Duke to put up £6000 for a glasshouse at Greenwich, where he would produce 'Christall de roache or Venice Christall' for ten years. By 1668 de la Cam had left England to form another glasshouse. In 1661 Martin Clifford and Thomas Powlden obtained a licence for their 'new invention of making christall glasses', which was converted to a patent in 1662 for fourteen years to the Duke of Buckingham for his agents, Clifford and Powlden. The patent was surrendered in 1663, but a further licence, covering mirror plate as well as drinking glasses, was granted to Thomas Tilson, a London merchant, in 1662. This was also converted into a patent in 1663, again to the Duke of Buckingham for fourteen years in the name of his agents, but the part relating to mirror plate was absorbed into the Duke's sole privilege for making mirror plate, granted to him in the same year. Buckingham's mirror glassworks, which was at Vauxhall, was managed from 1671–4 by an Englishman, John Bellingham. John Evelyn, who visited the Lambeth works in 1676, commented 'they make huge vases of mettal as cleare, ponderous and thick as chrystal: also looking-glasses far larger and better than any that come from Venice'.

George Ravenscroft and the quest for a new cristallo

The incorporation of the Royal Society in 1662 heralded a new age of experiment in the glass industry. A determined policy to find an alternative to Italian *cristallo* was instituted by the London Company of Glass Sellers when it was at last granted its charter of incorporation in 1664. (Charles I had sold it its charter twenty-six years previously, but the City of London had refused to implement it.)

Patents for new forms of *cristallo* were granted to Clifford and Powlden, to Tilson and finally to Ravenscroft between 1660 and 1674; and in 1663 Bryan Leigh, Adam Hare, William Broughes and Ralph Outlye applied for a patent for 'a way never yet before discovered, of extracting out of Flinte all Sorts of lookeing glasses, plates both Christall and ordinary and all manner of Christall glasse, farr exceedeing all former experiments both at home and abroad'. These patents were more variations on old themes than real inventions: potash was gradually being substituted for soda, and a better quality silica introduced, with oxide of lead probably used as a flux.

Dr Christopher Merrett's translation of Antonio Neri's standard Italian book on glassmaking, *L'Arte Vetraria* (1612), was completed by 1662, and it undoubtedly helped George Ravenscroft (1618–81), who was engaged by the Glass Sellers' Company in 1673 to carry out research into the making of glass in England. Son of a prosperous shipowner who traded with Venice, Ravenscroft was well versed in science, and initially had the help of Italian glassmakers, notably da Costa, an Altarist, who had a working knowledge of Italian glassmaking materials. At his experimental glasshouse in the Savoy, London, he set about trying to find 'a new sort of crystalline glass resembling rock crystal', and was granted a patent for this in 1673 for seven years. His work was impressive enough to encourage the Glass Sellers to set him up in a more secluded glasshouse at Henley-on-Thames in 1674.

His first experiments, using English flints and potash to replace Venetian pebbles and *barilla*, were not successful, since the glass tended to 'crissell' ('crizzle'). (Crisselling refers to the network of tiny cracks which appears in glass when its chemical structure breaks down because of an excess of alkali.) Ravenscroft remedied this by replacing a proportion of the potash with oxide of lead, which was eventually increased to as much as 30% of the mix. Sand was soon used to replace English flints which had been calcined and ground to a white powder, but the name 'flint glass' stuck until nearly the modern period. In 1676 it was announced that the defect of crisselling had been overcome, and it was arranged that Ravenscroft could use the seal of a raven's head as

Plate 29 English posset pot in clear colourless glass with a seal at the base of the spout, George Ravenscroft *ca* 1677. *Height 86 mm (3.38 in)*

his device on the Company's products in honour of his achievement. His glass of lead had a lustrous appearance and a brilliance which was to establish England as the leader in the production of clear, colourless glass from the end of the seventeenth century (*pl 29*).

Italian *cristallo* continued to be imported into England, the Glass Sellers' Company insisting on high quality from both English and Venetian manufacturers. The letters from Michael Measey and John Greene, Glass Sellers of the King's Arms, Poultry, with the firm of Allesio Morelli between 1667 and 1672, vividly illustrate the styles and quality in demand at the time. They reveal that Greene imported around two thousand dozen glasses and over one thousand looking-glass plates following the Restoration – an indication of the size of the Glass Sellers' activities. Although Greene and Measey asked for a 'verij cleer whit sound mettall', their letters reveal they were none too pleased with the quality and clarity of the glass they received, and probably turned to the home producers after 1672.

In 1678, Ravenscroft's agreement with the Glass Sellers was terminated, and in 1681, the year of his death, his patent expired. The

Glass Sellers took a lease on the Savoy glasshouse in 1681, and in the following year employed his former assistant, Hawly Bishop, to make 'christaline or flint glass'. In the meantime the company had signed an agreement in 1678 with two London firms, Michael Rackett in the Minories, and John Bowles and William Lillington in Stoney Street, Southwark, for the provision of 'white glasses'. The use of lead in glassmaking had probably reached the Continent by 1680, if not before, and by the time John Houghton wrote his letter, published 15 May 1696, giving a list of glasshouses in England, there were twenty-seven establishments engaged in the production of flint glass out of a total of eighty-eight (see p 198).

Leading London firms to adopt the lead glass after 1681 were Michael Rackett, John Bowles and William Lillington, the Duke of York's Glasshouse (Hermitage Stairs, Wapping), the Salisbury Court Glass Company, and Francis Jackson and John Straw, who had at least two glasshouses at the Falcon Stairs and one at King's Lynn. According to Houghton there were five flint glasshouses in Stourbridge, run by English rather than foreign glassmakers, such as the families of the Rogers, the Bradleys, the Grazebrooks, the Pidcocks and the Honeybournes, and the Penns and Hawkes of Dudley. Lead glass was introduced to Newcastle-on-Tyne about 1684 by the Dagnia family, and Bristol, about 1691, by John Perrott, John Little and others. By 1690 a Dublin glasshouse was producing lead glass in the English style.

Houghton's list shows that the main concentrations of glasshouses were in the London, Stourbridge, Tyneside and Bristol areas. Bristol was to remain a major glassmaking centre until after the eighteenth century, when its importance rapidly declined.

The years 1681–95 were a period of prosperity for the British glass industry, which was largely due to the efforts of the Glass Sellers' Company. The brilliance of English glass-of-lead established it as the leading product on the European glass market, and there was a healthy export trade by the end of the seventeenth century. Because of its success, the glass industry was subjected to a heavy tax, introduced as an emergency measure on 29 September 1695, and made perpetual the following year, in order to supply money for the French war. It was bitterly opposed by the industry for four years, until the tax was withdrawn on 1 August 1699. During this period, glassmakers had had to pay a 20% tax on flint glass, and one shilling per dozen on bottles.

In 1712 the Glass Sellers' Company was allowed by the City to increase its Livery from twenty assistants to sixty 'because the trade of glassmaking especially in Flint and Looking glasses is much improved of late years', appropriate recognition for its work during the previous fifty years.

Plate 30 Large goblet in clear colourless glass, made in England at the beginning of the eighteenth century. *Height 286 mm (11.25 in)*

At first English lead glass products were still subject to the Venetian influence which had dominated the market for over two centuries, but political and economic factors lead to a complete change in style by the end of the eighteenth century. The Treaty of Utrecht in 1713 allowed the influence of German glassmakers to reach England, and the accession of George I of Hanover to the English throne in 1714 brought many continental craftsmen to the country who introduced new styles. Until the Glass Excise Acts of 1745, 1777 and 1787,

Plate 31 Engraved English 'Excise' wine-glass with hollow stem, mid-eighteenth century. *Height 165 mm (6.5 in)*

English glasses tended to be heavy, simple and capacious, and were often sold according to their weight (*pl 30*). Since the Excise Acts taxed glasses on their weight, and not on their value, the inevitable result was that by the end of the century glasses were considerably lighter, and the glassmaker sought to lure his customers by decoration such as cutting and engraving, rather than by weight. The first Act came into operation in 1746, the tax was doubled in 1777, and increased in 1781 and 1787, and brought the industry great hardship. The duty continued to be raised step by step up to 1825, when the country's glass production, measured in weight, was down by one half. Opaque glasses were not included in the 1745 Act, but were in the 1777 Act, which may account for the popularity of opaque white enamelled and painted vessels, and opaque twists in stems of wine

Plate 32 Bowl and dish in clear colourless glass with cut decoration, Ireland, *ca* 1825. *Length 222 mm (8.75 in)*

glass around this time. (*pl 33*) The so-called 'Excise' glasses of the eighteenth century had hollow stems to take away some of the weight (*pl 31*).

Eighteenth-century drinking glasses are among the most popular items sought after by collectors today. They include cordial, dram, firing, rummers, champagne, strong-ale and toast-master types. Stems may be plain, baluster, incised twist, air and enamel twist, or cut, and decoration of the bowls could be plain, moulded, engraved, cut, gilded or enamelled. Besides drinking glasses, common articles included serving bottles, decanters, squares, carafes, sweetmeat glasses, salvers, wineglass coolers, finger bowls, candlesticks, taper sticks, girandoles, tumblers, fruit and salad bowls and tea jars.

It is impossible to cover here the wide variety of popular shapes and decorations prevalent in eighteenth-century table glasses, since it is a huge subject on its own. Many books have been written about it, but a good grounding can be gained from Ernest Barrington Haynes'

Glass Through the Ages, or George Bernard Hughes' *English, Scottish and Irish Table Glass*.

The Excise Acts, combined with the full free trade rights granted to Ireland in 1780, led to a considerable expansion in glassmaking there, notably at Cork and Waterford, where cut glass virtually indistinguishable from the English product was made. (*pl 32*).

Despite a considerable overseas trade, the British glass industry was suffering badly between the years 1785–1835 through the successive Excises, which were rigidly enforced. In 1788 a glasshouse at Nailsea

Plate 33 English tea jar in opaque white glass with enamelled decoration, South Staffordshire, *ca* 1770. *Height 143 mm (5.63 in)*

was founded by a Bristol bottle-maker named John Robert Lucas, who was reputed to have avoided paying some of his Excise duty by making common domestic vessels in green bottle glass (*pl 34*). The most noteworthy period of the Nailsea glasshouse, which made mainly window glass, came under the management of Robert Lucas Chance between 1810 and 1815, who took over the famous glasshouse at Spon Lane in Birmingham in 1824. Another glassworks was established at Holland Street, Southwark, about 1790 by Apsley Pellatt I. His Falcon Glasshouse continued to make glass until 1878, and he also made glass at New Cross until 1895. But it was the founder's son,

Plate 34 English jug in bottle-green glass with small pieces of opaque white glass marvered-in, Nailsea-type, early nineteenth century. *Height 181 mm (7.12 in)*

Apsley Pellatt II (1791–1863) who made the name famous. He acquired a wide knowledge of glassmaking techniques from all over Europe and was the first Englishman to make a study of ancient glassmaking processes. He published two books on glassmaking in 1821 and 1849. In the latter, *Curiosities of Glassmaking*, he described several Venetian techniques, including 'ice' glass, and the process of cameo incrustations or 'crystallo ceramie' which made his firm famous (*pl 35*).

Plate 35 Perfume bottle in clear colourless glass with cut decoration and panels in the *crystallo ceramie* technique, made by Apsley Pellatt at the Falcon Street Glasshouse, Southwark, London, after 1820. *Height 140 mm (5.5 in)*

Plate 36 Pair of tumblers in clear colourless glass, press-moulded to imitate cut glass, England, 1850. *Height 108 mm (4.25 in)*

Nineteenth-century British glassmaking

The arrival of press-moulding from America in about 1830 allowed glass to grace the tables of the poorest British households. The pressing mechanism, which consisted of a table fitted with interchangeable metal moulds, and a plunger which thrust into the mould from above, produced ready-decorated tableware, often with imitation cutting, without involving blowing or manipulation of any kind (*pl 36*). Pressed glass was made in England notably by Henry Greener in Sunderland, and George Davidson's and Sowerby's at Gateshead.

In 1833 there was sufficient cause for concern about the glass industry for a Royal Commission to be appointed to assess the state of glassmaking and the disastrous effects of the Excise Acts on the administration of glasshouses and glass production. Nothing was done for twelve years, but in 1845, exactly a century after the first one came into force, the Acts were repealed. The result was to stimulate interest in glass technology, culminating in the extraordinary exhibits from glassmakers in the Great Exhibition of 1851. The firm of F. and C. Osler of Birmingham included in its display a glass fountain weighing four tons, and the Venetian-style glasses of Pellatt were also exhibited. Anticipating the opulence of Victorian glass for the next half century were the glasses of George Bacchus & Sons of Birmingham, the coloured glasses and cut ruby casings of Rice Harris & Company of Birmingham and Davis Greathead and Green of Stourbridge, and the cut and coloured glasses of W. H. P. Richardson of Stourbridge.

The Great Exhibition marked the beginning of a new era in British glassmaking. Its most remarkable result was a museum-taught revival of the ancient techniques of glassmaking. Old Venetian styles were reintroduced but were often over-elaborate in execution. Bohemian styles in coloured and cased glass, gilt designs in 'Arab' style, and medieval Syrian enamelling also reappeared, adding to the impact of Japanese art which was felt between 1875 and 1885. Cameo glass, harking back to Roman times, was made by several Stourbridge firms from the designs of John Northwood, George and T. Woodall, Alphonse Lechevrel and others. Stourbridge was a great centre for glassmaking during the nineteenth century and contributed immensely to the development of the industry. The four oldest Stourbridge firms were Richardson's of Wordsley, Thomas Webb & Sons of Stourbridge, Stevens and Williams of Brierley Hill and Stuart's Red House Glass Works in Wordsley.

After the 1851 Exhibition, cut glass went out of fashion. However, the technique was never totally abandoned, coming internationally to the fore once more in the 1880s and '90s, and it has remained a status symbol even to the present day. Victorian glassmakers were in hot competition with manufacturers on the continent and in the U.S.A. in the second half of the nineteenth century to produce new types of art glass. New methods of decorating and colouring glass were regularly introduced by glass technologists such as Joseph Locke, who was also a painter, engraver, etcher, sculptor and inventor. Another post-Exhibition development was the introduction of outside designers, not necessarily glassmakers, to develop new glass shapes. In 1859, Philip Webb, an architect employed by William Morris, designed wine glasses and tumblers, which were made by James Powell and Sons at the Whitefriars glassworks in London. The same firm produced glasses for Morris to the design of another architect, Thomas G. Jackson, In 1874. The criteria for these glasses were that the glass should look soft and should attain its design from pure furnace work, and they influenced the design of English glasses for the next sixty years.

The bottle industry

Fragments of open-mouthed, bag-shaped bottles in pale green metal were common among the finds from mid-fifteenth-century levels during post-war excavations in London. By the end of the fifteenth century large quantities of bottles, with pushed up bases, were being produced in the Weald. Multi-sided moulds were apparently used by Carré's Lorrainers in England in the 1590s, but the use of moulds in bottle-making seems to have died out in the mid-seventeenth century.[18]

Plate 37 Dark green glass bottle with seal, made in England by H. Ricketts in the first half of the nineteenth century. *Height 235 mm (9.25 in)*

Archaeological evidence does not indicate that wine bottles were made much before 1650, but a bottle industry grew rapidly in the second half of the seventeenth century, quite distinct from other branches of glassmaking. By 1696, Houghton could list around forty-two bottle houses, producing nearly three million bottles annually, and could find records for the importing of only eight dozen bottles from Sweden. There was little change in the techniques of bottle-making during the seventeenth and eighteenth centuries, the glassmakers relying on their basic equipment of blowing iron, mould, marver and pontil rod (or 'punty' – a solid iron rod, which when removed from the glass, left the familiar rough 'pontil mark' at the base of the vessel).

Seals appeared on bottles from the mid-seventeenth century, made from a blob of hot glass dropped on to the side of the bottle and stamped with name, initials, rebus, arms or sign of the owner, which could be an individual, a tavern or other institution. The earliest dated seal ever found – and lost again – bore the initials 'C. B. K.' and the date '1562', but the earliest surviving seal is that of John Jefferson dated 1652.

A major change came in 1811, when Jacob Wilcox Ricketts and his son, Henry, of Bristol, perfected a mechanical method of producing bottles. Metal moulds shaped the whole vessel, including the string rim, in a single operation (*pl 37*). In 1880 another Englishman, William Ashley, invented a semi-automatic bottle-making machine, but the first fully automatic bottle-making machine was built in Toledo, Ohio, by an American, Michael Owens, in 1903. The huge difference that these machines made to the industry can be measured by comparing the 1696 annual total of 2,880,000 containers made in Britain with the present total of around fourteen million a day.

Twentieth-century developments in British glassmaking

During this century, though new techniques have still greatly interested glassmakers, art glass has reflected the inherent qualities of the material, rather than added decorative effects. The twentieth-century trends can be divided into three main movements: craft, engraving and stained glass. It is impossible to do any justice to the subject here, so just a brief mention of a few names will be made.

The architect Keith Murray is well known for the simple, sometimes facet-cut vessels he designed for Stevens & Williams at Stourbridge from 1932. James Hogan (1883–1948), who became director of Whitefriars (James Powell & Sons), created a new style for blown glass which received international acclaim. His plain well-proportioned vessels occasionally had furnace-worked decoration, but cut or engraved decoration was rare. Geoffrey Baxter (b. 1922) became resident designer at Whitefriars in 1954. He absorbed Hogan's style of work, giving it his own discreetly personal expression, especially in large, furnace-worked vases.

It was Whitefriars glass which was used by the New Zealander John Hutton, when he translated some of his famous engraved figures in Coventry Cathedral on to large vessels. Hutton's distinctive shallow light engraving with its rough finish can also be seen at Guildford. Other glass engravers highly acclaimed in Britain include Laurence Whistler, Jane Webster and David Peace. Among the most outstanding stained glass works are the wonderful designs of Patrick Reyntiens and John Piper which crown the new Roman Catholic Cathedral in Liverpool.

Old techniques are still very popular, especially cut glass, with excellent quality products still being produced by old established firms such as Stevens and Williams, Royal Brierley Crystal, and Stuart and Sons Ltd. Glassmaking in Britain is still as hugely competitive as ever, with new technological developments creating new markets all the time.

1. *Hughes 1972*, 99
2. *Charleston 1963*
3. *Harden 1961*; *Atkinson 1929*; *Wilson 1974*
4. *Alcock 1963*, 187
5. *Harden 1956*
6. *Harden 1961*
7. *Simms 1894*, 3
8. *Cramp 1970, 1975*
9. See *Kenyon 1967* for Wealden glassmaking
10. *Neale 1974*
11. *Godfrey 1975*, 24–25
12. *ibid*, 35
13. See *Godfrey 1975* for details on monopolies
14. *Kenyon 1967*, 52
15. *ibid*, 47
16. *Godfrey 1975*, 191
17. As 13
18. *Hume 1961*, 92

CHAPTER SIX

Archaeological Evidence in Britain

Little archaeological evidence has been found to indicate what British glass furnaces of the Roman period looked like, or how they operated. Excavations of Roman glasshouse sites found at Wilderspool, Cheshire; Middlewich, Cheshire; Mancetter, Warwickshire; Wroxeter, Salop; and Caistor St Edmund, Norfolk, have revealed little structural detail.

The five workshops found at Wilderspool were excavated and described by Thomas May at the beginning of this century. He describes the furnaces as small oval ovens with outlets and flues.[1]

More recent work on a Roman glasshouse site by Katherine F. Hartley at Mancetter, Warwickshire, in 1964–5 and 1969–71, uncovered a similarly small and almost circular furnace. She describes her discovery as follows:

> Our glass furnace is in Mancetter parish, but it is in the midst of an industrial site of the Romano-British period (SP326967) which is linked with the Romano-British roadside settlement at *Manduessedum* on the Watling Street. The industrial area was primarily a pottery-making site which almost certainly stretches from *Manduessedum* to Hartshill, two miles away. It seems likely that there would be other glass furnaces, but it is so small that it would be quite easy to miss them as we have excavated only a very small area of the whole site.
>
> The first version of the furnace was 2 feet 6 inches (80 cm) long and 2 feet 5 inches (77 cm) wide – virtually circular. It was relined or patched at least three times, and the dimensions of the final version, whose lining was well preserved, was 1 foot 8 inches (51 cm) long and 1 foot 1½ inches (34 cm) wide. The tiles in the bottom are all broken ones which were put in for the last phase. Their surface has not been reduced where it underlies the extra clay put in for that version. The original floor was probably a clay-lined one. We thought that the shale and stone foundation connected with it was probably for the annealing oven. It was clear

that cullet was being used here, but we found many small fragments of waste glass and there was some solidified glass surviving in a dribble on the final lining.

The furnace is in the middle of an area of intense activity and is undoubtedly not earlier than the mid-second century and could well be sometime later. There was a rough stone working floor nearby, which could have been associated with the furnace. We did find large numbers of fragments from the clay-lining in the filling, suggesting that it was a good deal higher originally. The trickle of solidified glass is the only sign of glass on the inner surface, and if the furnace had itself contained the molten glass, it would surely have left some deposit. Despite extensive excavations, we did not find any trace of a glassmaking pot. There is every reason to believe that the furnace was demolished and filled in in the Roman period (*pl 38*).

Plate 38 Phase II of the Roman glass furnace found at Mancetter, Warwickshire, linked with the Romano-British roadside settlement at Manduessedum on the Watling Street. It was established not earlier than the mid-second century AD. The broken tiles in the bottom date from the last phase of the furnace, which was demolished in the Roman period.

Dorothy Charlesworth of the Department of the Environment, who was working on the glass from the excavation, described it as natural blue-green glass with drips and threads in reasonable quantity to attest manufacture. She agreed that cullet, or broken glass, must have been used to make the glass, since she did not see how raw materials could have been used in such a small, isolated furnace. Analysis showed the glass to be of a soda-lime type.

Although so few Romano-British glasshouses have been discovered, it seems likely that any major site of this period would reveal at least one glass furnace making utility glass, even if only window glass, for local consumption.

The ground plan of the Saxon glassworking found at Glastonbury appears to have been oval, rather than rectangular. The coloured window glass which was produced at Glastonbury was probably associated with the period of the monastic revival in the tenth century (see p 63).

We have far more evidence about medieval British glasshouses. The greatest concentration of excavated sites is in the Weald, where more than forty sites are known, dating from *ca* 1330 to 1618. Nearly all are in six or seven adjacent parishes centred around Chiddingfold,

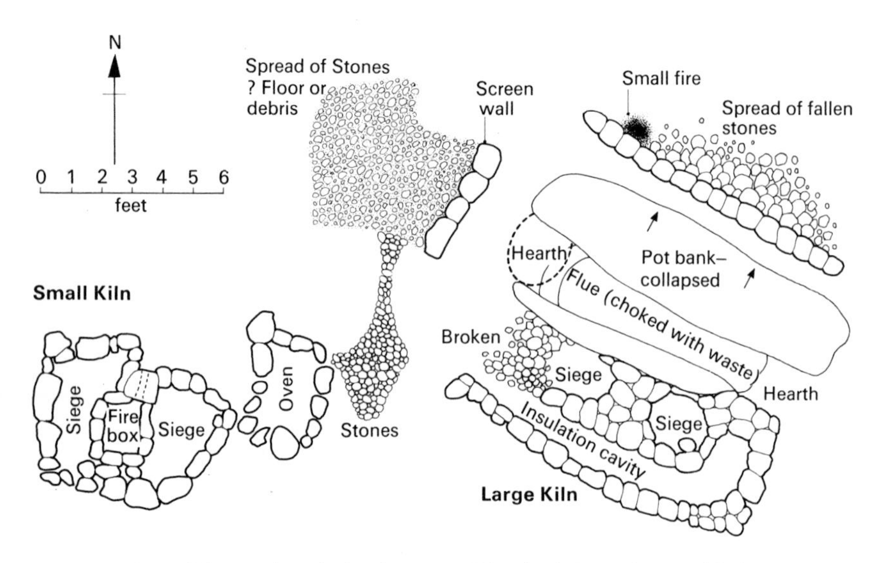

Fig 14 Plan of the medieval glasshouse at Blunden's Wood, Hambledon, Surrey.

Plate 39 Reconstruction of a medieval glasshouse, based on the excavation at Blunden's Wood of a site dating from approximately 1330. Note the main furnace with separate smaller furnaces to the left of the picture.

Kirdford and Wisborough Green. Others no doubt remain to be discovered.

Kenyon divides the Wealden industry into two phases, Early (1330–1567), when methods were primitive, with poor quality glass and a limited range of products, and Late (1567–1618), where methods, quality and range showed a noticeable improvement. Furnaces from the Early period were generally constructed from stone embedded in clay, but in the Late period, brick was often used for the whole or part of the structure.

The earliest known Weald site is that at Blunden's Wood, Hambledon, Surrey, excavated in 1960 by Eric Wood. Associated pottery, supported by magnetic dating, indicated that the furnaces here belonged to the second quarter of the fourteenth century, probably *ca* 1330. Blunden's Wood glasshouse followed the usual forest glasshouse plan, the main furnace being roughly rectangular in shape, with two parallel siege-banks, and a fuel trough between them. It was a wood-fired furnace, eleven feet long and originally eight feet wide, holding four pots, two on each siege. Only two feet in height of the original structure remained, and no real evidence of the nature of the furnace roof was found (*fig 14*).

Two small kilns were found within seven feet of the main furnace, the larger round kiln having a central fire box. Wood suggested that the left side of the round oven was used for fritting, and the right for annealing. The small oven, he conjectured, was probably used for preheating pots. There were no foundations to the furnaces, which were constructed from local sandstone, the sieges built up in dry-wall construction.[2] A reconstruction of a medieval glasshouse based on the evidence from the Blunden's Wood excavation is on display at the Pilkington Glass Museum, St Helens, Merseyside (*pl 39*).

A Weald glasshouse dated magnetically and by the discovery of a coin at the site to the early 1550s was excavated at Knightons, Alfold, Surrey by Eric Wood in 1973 (*fig 15*). The glass found at the site was distinctly of the early Weald type. There were four furnaces, the first

Plate 40 Medieval hanging lamp of Weald-type glass, found on the site of the new county council offices, Winchester, 1955.

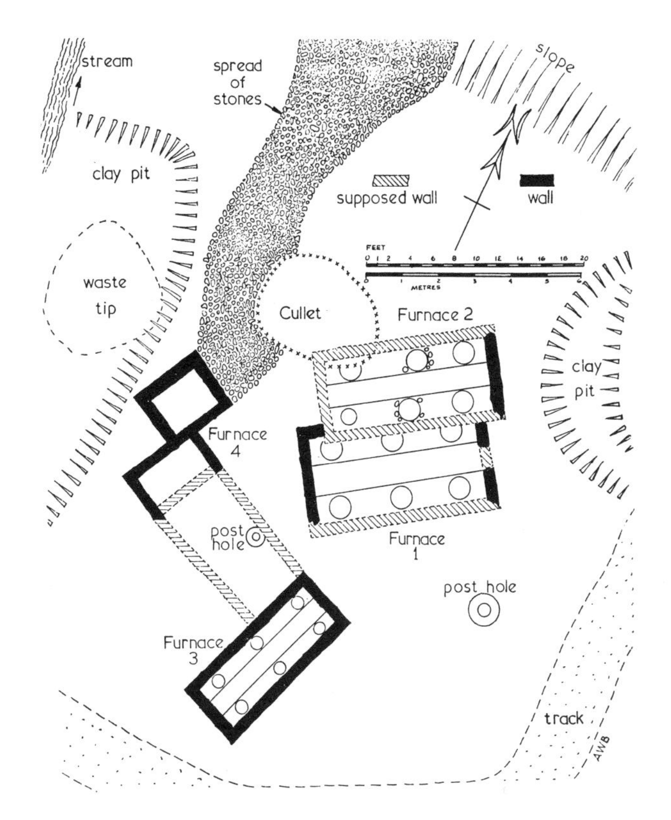

Fig 15 Plan of the *ca* 1550 glasshouse at Knightons, Alfold, Surrey. Note the three six-pot furnaces, furnace 2 overlying furnace 1, and the two-chamber annealing furnace 4.

of standard Wealden type – rectangular, with banks for three pots on each side. The second furnace, which was of the same design, overlay the first, and hence must have replaced it. The third, built nearby, similarly held six pots and probably replaced the second furnace. It was connected with, and no doubt fed, a two-chamber annealing furnace designed to take crown window glass sheets, the first example of crown glass manufacture in England.[3]

Knightons provides a vivid illustration of the short life of these early furnaces. Warren C. Scoville remarks that extensive repairs were often necessary for a furnace several times a year, and the life-expectancy for a furnace was seldom longer than two to three years.[4] Furnaces which were operated on a seasonal basis, such as those run by farmers like the Strudwicks and Peytowes in the Weald, may have had a slightly longer life. Their furnaces shut down for two or three months in summer, probably to fit in with the farming cycle. The possibility of two main furnaces existing on the same site, as at Knightons, has been recorded on other sites such as Malham Ashfold, Wisborough and Chaleshurst, Chiddingfold.

Crucible fragments found at the numerous Wealden glassmaking sites have shown that there was no particular shape that was in general use, and crucibles could be large, small, straight-rimmed, flared, inturned-rimmed, straight-sided or funnel-shaped. The only common factor was that they were all open.

Documentary and archaeological evidence indicates that the working areas of the glasshouse were protected from the weather by a roof or shed of simple wooden construction. During the Middle Ages wooden shingles seem to have been used for roofs more commonly than thatch, with tiles replacing the shingles by the sixteenth century. Tiles with peg-holes were found at Knightons.

The Bagot's Park, Staffordshire, glasshouse, excavated by David Crossley in 1966, gave the first clear evidence of an all-over timber and tile roof covering the furnace area. Fifteen glassmaking sites were found at Bagot's Park during land reclamation, but only one, dated to the early sixteenth century, was excavated. It revealed a six-pot furnace, with clear indications of a clay roof built over the sieges and the fire box. The roof had been formed with a stiffening of twigs which burnt out as the clay hardened, after the fire in the two stoke holes at either end of the furnace had been lit. It is assumed that there were six 'gathering holes' in the sides of the furnace, which may have been used to allow fumes to escape. (No evidence has been found at any site to prove that there were chimneys or gaps in the tops of early furnaces for this purpose.) Ditches provided drainage, and at each end of the furnace there was a pair of holes capable of taking posts fifteen inches in diameter. Stone packing in the holes probably helped to support an all-over roof which was covered by tiles found in adjacent rubbish. A small stone and brick furnace built close to the main furnace was probably used for annealing.[5]

Since water was necessary for washing raw materials and cooling purposes, furnaces were generally built near rivers, streams, or ponds. Drainage ditches have been found surrounding the furnaces at Fromes Copse, Chiddingfold, Wephurst, Kirdford, Malham Ashfold, Wisborough and – the best example – at Hazelbridge Hanger, Chiddingfold.

Traces of glassmakers' dwellings have yet to be definitely located on a glasshouse site. Pottery found in some abundance on sites like Blunden's Wood, Bagot's Park, and Hutton and Rosedale, in North

Yorkshire, consists mostly of jugs, cooking vessels, storage jars, dishes and bowls, which indicates that the glassmakers certainly ate near the glasshouse, but did not apparently live on top of their work.

The available evidence suggests that the basic structure of the English wood-fired furnace varied little between the early fourteenth and late sixteenth centuries. After this, however, in the period which Kenyon calls the Late phase in the Weald (1567–1618) there were changes in design and in production methods. Bishop's Wood, Staffordshire (*ca* 1584–1604), for instance, one of six sites in the area, is more compact than the traditional glasshouse. Wealden sites of the Late period which have revealed interesting structural remains include Fernfold, Wisborough Green and Vann Copse, Hambledon. The Fernfold site is probably where Carré's glasshouse at 'Fernefol Sussex' stood *ca* 1567. Its furnace was built with rigid

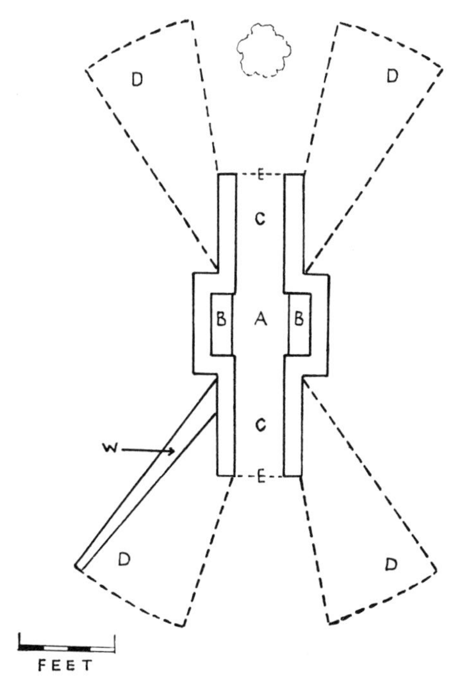

Fig 16 Plan of the Vann Copse furnace at Hambledon, showing four 'wing' furnaces attached to the main furnace. A = fire chamber; B = sieges; C = hearth; D = ?annealing chambers; E = hearth lip (tease hole); W = taper wing wall. Based on S. E. Winbolt's rough sketch modified by A. D. R. Caroe.

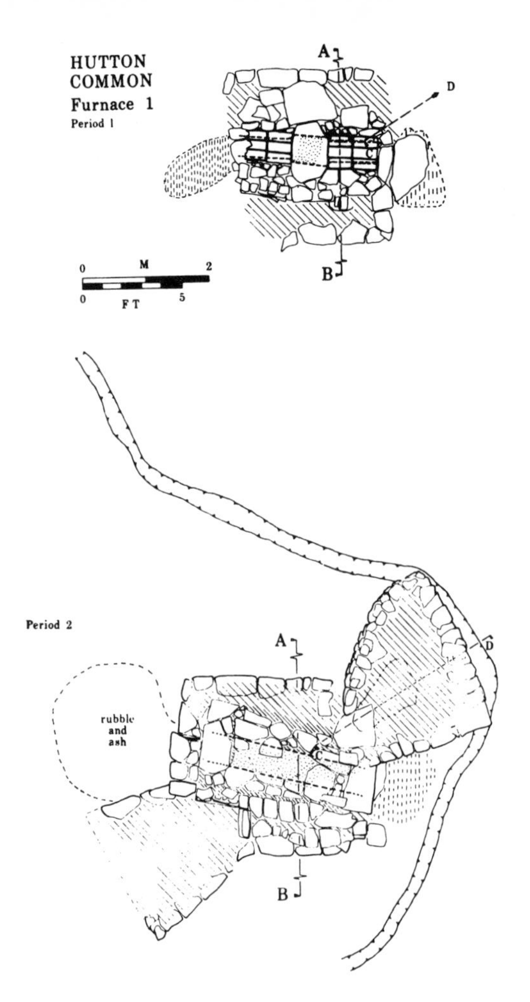

Fig 17 Plan of the sixteenth-century glasshouse at Hutton Common, North Yorkshire, showing two diagonally opposed 'wing' furnaces attached to the main furnace. Letters refer to sections drawn during excavations.

geometric lines to the northern pattern, with parallel sieges, and was constructed of 'well laid' bricks and sandstone. Like the Vann Furnace (probably late sixteenth century) which was also constructed of red brick, the Fernfold furnace could be stoked from both ends. Sidney Wood, Alfold, the site of a glasshouse built to a rectangular plan, may have been the last forest glasshouse to be worked in the early seventeenth century. The window and vessel glass produced

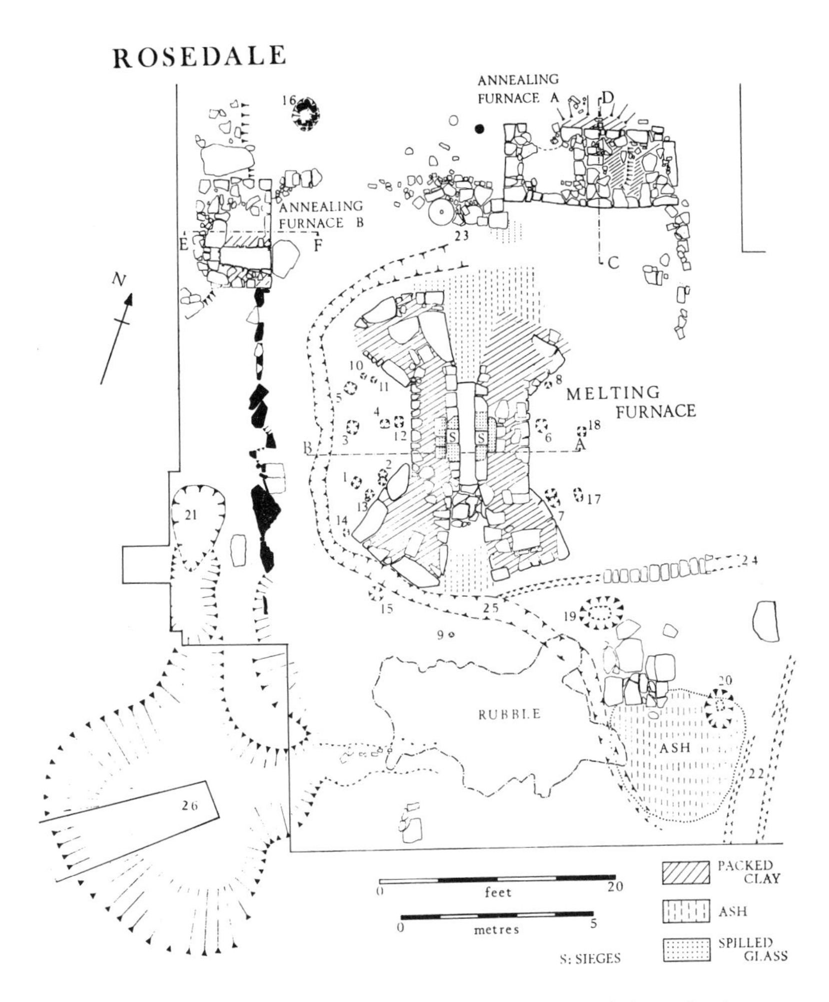

Fig 18 Plan of the sixteenth-century Rosedale, North Yorkshire glasshouse showing four 'wing' furnaces attached to the main one. Letters refer to sections drawn during excavations; numbers refer to postholes and other features such as drainage ditches and pits.

there is the finest that has been found in the Weald. Coal cinders were found here as well as at the Somersbury, Ewhurst (*ca* 1567+) and Petworth Park, Lurgashall (early seventeenth century) sites, indicating that some experimentation with coal fuel was going on in the Weald at this time.

Towards the end of the sixteenth century wedge-shaped 'wing' furnaces began to be added to the main furnace of English glasshouses, the first known example being at Buckholt, Hants. (1567–9). The irregular, detached walls, which the excavator called 'rays', were made of flint, and appeared to radiate away from each corner of the small oblong furnace. The Vann Copse furnace was built in the traditional rectangular plan, but with four short wings running diagonally from each corner (*fig 16*).

The excavations of furnaces at Hutton and Rosedale, North Yorkshire, by D. W. Crossley and F. A. Aberg between 1967 and 1971, revealed two more winged furnaces dating to the late sixteenth century. The main furnace at Hutton, in its later 'experimental' phase, had two wings constructed against the north-east and south-west corners of the central block. Traces were found of a second furnace, about three metres to the west; this was possibly an annealing oven built during the early phase of the glasshouse, before the wings were added (*fig 17*).

Ten kilometres to the north-east of this site, in the Rosedale valley, a further glasshouse was found. It had been making similar vessel glass and was apparently run by the same glassmakers. As the Rosedale furnace was superior in quality to the one at Hutton in every respect, including crucibles and glass, it seems reasonable to suppose that the glassmakers built this furnace after the Hutton enterprise was established. The most interesting thing about the rectangular two-pot furnace at Rosedale is that it had four wing kilns, built at much the same time as the central block of the furnace, magnetically dated to 1580–1600. This is therefore the most sophisticated wood-fired glass furnace to have been discovered in England. Two additional furnaces, probably for annealing, were found near the main furnace, and drainage ditches and post holes were found on both North Yorkshire sites. The wing furnaces would have been used for fritting, annealing or pre-heating crucibles, probably drawing their heat from the main stokeholes through the main melting furnace (*fig 18*).[6]

Although main furnaces continued traditionally to be rectangular in shape, we have one example of a possible round main furnace, found at Woodchester, Gloucestershire, around 1880. The furnace was sixteen feet (4.8 m) in diameter, was surmised to be beehive-shaped and had a rectangular annealing oven nine feet by seven feet (2.7 m by 2.1 m) nearby. It lay in an enclosure fifty feet (15 m) square, with a wall on one side and possibly a lean-to shed over it. The good quality glass, similar in all respects to the products of other glasshouses, indicated a 1590–1615 date.[7] Forest glasshouse sites which have been located in Britain in recent years include Ruyton-XI-Towns, Salop, found by Michael Roe in 1967 on a meadow bordered by the River Perry; Biddulph, Staffordshire, located by the Biddulph Historical Society in 1970 adjacent to a garage on the Congleton-Biddulph road; Glasshouse Farm, Staffordshire, investigated by John Gask of the Keele and Newcastle Archaeological Society in the Bishop's Wood area in 1968, which turned out to be a glasshouse dump; Glasshouse Wood, Ashow, near Kenilworth, investigated by Ray Wallwork and Eric Willacy of the Coventry and District Archaeological Society in 1970–71; Buriton, Hampshire, found in 1971 and excavated by the Reverend Peter Gallup, Elizabeth Lewis (Winchester Museum), John Budden and others in 1973, with a second site located in the same area of Glass Brow, Ditcham Woods, in 1976; Tugley Wood and Broadlands, Ramsnest Common, Chiddingfold, Surrey; two sites at Shillinglee Park, Plaistow, West Sussex; and Bickerstaffe, near Ormskirk, Lancashire, located and excavated by the author in 1968–69.

Even before James I's Proclamation of 1615 forbidding the use of wood fuel the design of glass furnaces had begun to be modified for using coal. The sudden need for a much larger draught through the furnace meant glassmakers were forced to think of ways of providing an under-draught to allow the coal to burn properly. Mention of a 'wind furnace' used to burn coal at the Winchester glasshouse in Southwark was made as early as 1612 in Simon Sturtevant's *Metallica*.

The first early coal-fired glasshouse to be located and excavated in England was found in 1968 at Haughton Green in Denton, near

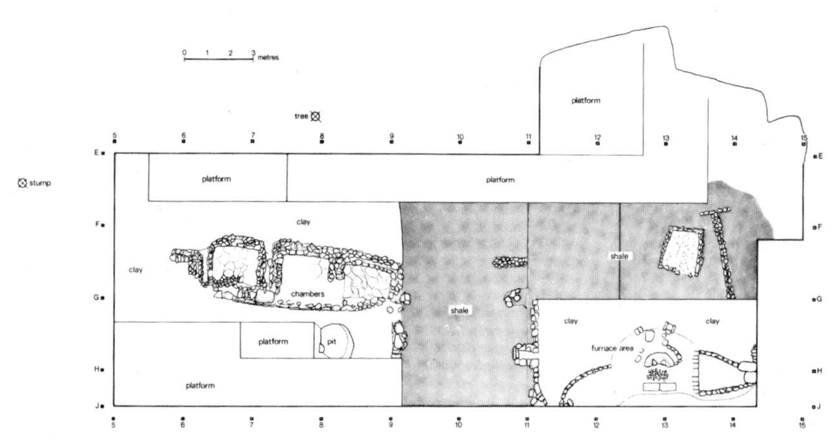

Fig 19 Plan of the seventeenth-century glasshouse at Haughton Green, Denton, near Manchester, after the 1970 excavations. ?Annealing chambers lie to the left, with the main furnace bottom right.

Plate 41 Annealing chambers at the Haughton Green glasshouse, Denton, near Manchester.

Manchester, during the laying of a new sewer line along the valley of the River Tame. Excavations carried out between 1969 and 1972 by the author and Freda J. Burke for the Pilkington Glass Museum revealed three well-preserved structures, probably used for annealing, and the main glassmaking furnace approximately six metres further up the river valley (*fig 19, pls 41, 42*). The main feature of the (?) four-pot furnace was a large flue system running beneath the two parallel sieges. The glasshouse produced fine quality black, blue, amber and green vessel glass as well as cylinder (broad) window glass, bottles and other more specialized items (*pl 43*).

As deposits lay up to three metres deep it was not possible to excavate the site fully, but stone steps were found leading down into the south end of the flue, which was more than two metres deep. (*pl 44*). The north end of the flue was shorter than the south end and was also built in rough stone packed with a little clay and straw. Dressed stone had been used to construct the arch over the wind tunnel, adjacent to the north end of the melting furnace, and the collapsed remains of a similar arch were found inside the south flue. The arch or barrel vault did not appear to extend to the end of the flue, but left a gap at either end. The flue curved slightly inwards towards the river. No traces of the iron grill which must have rested in the space between the two sieges were found, but the well-made and much used stone steps into the south end of the flue were no doubt used for scraping out the large amounts of ash and cinders left by coal burning. The back of the stone-built siege banks was packed with red-burnt clay, and there were the remains of an outer stone wall. A large amount of heavily burnt furnace debris behind the siege banks, including the remains of other sieges, bore witness to the much hotter conditions the Haughton Green glassmen had to contend with. A gathering hole cover was found, similar to one found at Bagot's Park (*pl 45*).

Plate 42 Aerial view of the main coal-fired furnace at the Haughton Green glasshouse. Note the central siege banks with existing flue archway to the left and partially excavated flues to left and right.

The Haughton Green glasshouse was worked by two well-known glassmaking families, the du Houx and the Pillmeys. References to them and to other glassmakers occur in the parish registers of Stockport, Ashton-under-Lyne and elsewhere between 1605 and 1653. Since there is no record of glassmakers after that date it is not improbable that the glasshouse was destroyed in one of the many Civil War skirmishes that took place in the district.[8]

More light will be thrown on the construction of the transitional coal-fired furnaces with the excavation of the Kimmeridge Bay glasshouse on the Isle of Purbeck, Dorset, whose location was reported by Philip Whatmoor in *BP News*, February 1976. Plans to excavate the site are now under way. This glasshouse, which was operated by Abraham Vigo in partnership with the landowner, Sir William Clavell, used the local Kimmeridge coal, but owing to quarrels between the two men, it existed only for a short period, between 1617 and 1623. A trial trench revealed the stone structure of a glass furnace with much oil shale ash, broken crucible and many dark green glass vessel fragments, principally from bottles.

A seventeenth-century site of a possible coal-fired glasshouse was located by T. Pape around 1930, on Waterhays Farm, off Red Street, Chesterton, Newcastle, Staffordshire. Excavation of the site by the Keele Archaeological Group commenced in 1977 with large crucibles measuring up to one metre in diameter, and a siege around one and a half metres across being discovered, as well as window glass and bottles in shiny green glass. The main furnace structure has yet to be located.

A rectangular glasshouse with a complex system of flues dating to the sixteenth and seventeenth centuries was located by Denis Ashurst at Bolsterstone, Stockbridge, South Yorkshire, but full excavation has not yet been possible.

The most important feature of the coal-fired furnace was the iron bars which formed a grill on which the coal could burn. Christopher Merrett described them in 1662 as follows: 'Sleepers are the great iron bars crossing smaller ones which hinder the passing of the coals, but give passage to the descent of the ashes.' The grill system is clearly shown in a drawing of a London glasshouse made by a Swedish architect in 1777–78.[9]

There were several new developments in glassmaking crucibles in the second half of the seventeenth century. Merrett describes pots which were twenty inches (50 mm) wide at the rim and narrowed down towards the base, as well as 'piling pots' which were set on top of larger crucibles and used for melting special coloured glasses.

Plate 43 Selection of typical finds from the Haughton Green site in opaque black, clear blue and green-tinted glass.

Although there is no historical or archaeological evidence, it is generally thought that covered crucibles were introduced soon after Ravenscroft's development of lead glass in the 1670s, since it was essential to protect the new glass from smoke during manufacture. The covered or 'caped' pots were constructed all in one piece.

In the late seventeenth century British glassmakers began to house their furnaces in cone-shaped buildings. Because the cone could have all its outlets closed while leaving the large underground flues open, a tremendous through-draught could be created which greatly increased the efficiency of the furnace. It is not certain exactly when the cone system first came into use, but Captain Philip Roche was building a cone in Dublin as early as 1696.[10]

A remarkably detailed pictorial description of the English glasshouse of the eighteenth century is given in Diderot and D'Alembert's French encyclopedia (1751–71). The illustrations of the '*verrerie Angloise*' show a typical cone glasshouse with a central four-pot furnace with four wing furnaces attached diagonally to each corner of the main furnace. An iron grating for the coal fuel lies between the sieges and a very deep flue system runs beneath the furnace, bringing in air from three outside points (*figs 20, 21, 22*).

The excavation of the Ballycastle glass furnace, Co. Antrim, Northern Ireland, by Gavin Bowie in 1974 revealed a plan reminiscent of the Diderot drawings. His preliminary report states that the glasshouse, which operated from its construction between 1753–55 until about 1771, was of the '*verrerie Angloise*' type. The foundations of the main circular wall were found, about 110 cm in width, its inside

Plate 44 Worn steps leading down into the south end of the flue running under the main furnace at the Haughton Green glasshouse.

Plate 45 A gathering hole cover used to stop up the apertures in the main furnace when not in use, found at the Haughton Green glasshouse.

diameter being estimated to be eighteen metres, designed to support a large brick cone structure about twenty-seven metres in height. The foundations of an annealing house were located immediately adjacent to the base of a main archway, through the main circular wall, and hence the oven within its outer wall, linked directly with the central glass-working area.

Despite wave and river erosion, which had worn away half the glasshouse, and the intrusion of World War II fortifications, sufficient evidence for the main flue was found, the location of a main entrance arch at working floor level being immediately above that of the main flue. Two shallow longitudinal brick arches were found, one in each side of the retaining walls of the ashpan, and a single remaining cross-piece of wrought iron, located some 70 cm above the centre of the two longitudinal retaining arches, survived to indicate the level of the firegrate. Immediately inside the main circular wall and leading into one side of the flue was a brick-lined chute, which was probably used for the disposal of furnace waste. This was a pot furnace, which used the local coal and produced at least four types of bottle, crown window glass and a fine glass flagon.[11]

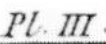

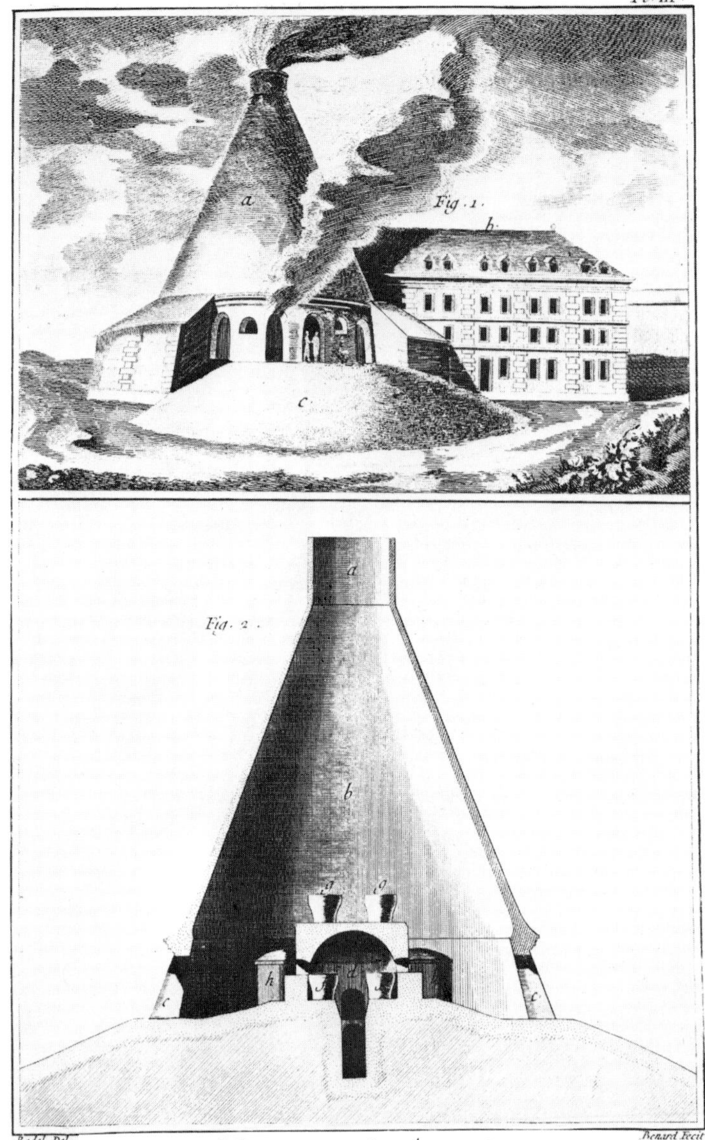

Verrerie Angloise.

Vue extérieure de la Verrerie, et Coupe sur la largeur.

Fig 20 Exterior view and section of a '*verrerie Angloise*' – an English-style cone glasshouse – from Diderot and D'Alembert's encyclopaedia. a, b = interior 'chimney' of cone building; f = crucibles on sieges; g = crucibles drying off on roof of furnace; h = annealing arches.

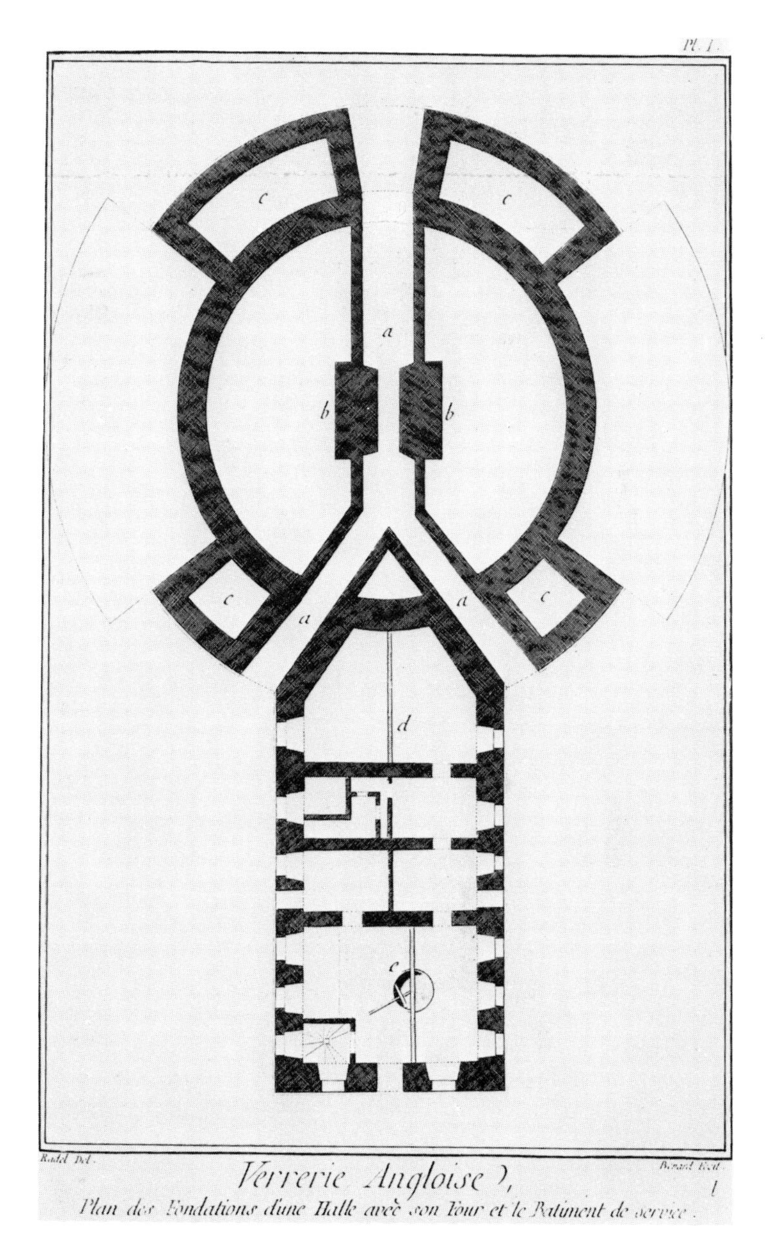

Fig 21 Plan of the foundations of an English cone glasshouse, showing the three large flues (a) which feed air from the outside to the central coal-fired furnace. (b) indicates the siege foundations, and (c) the annealing furnace foundations. From Diderot and D'Alembert's encyclopaedia.

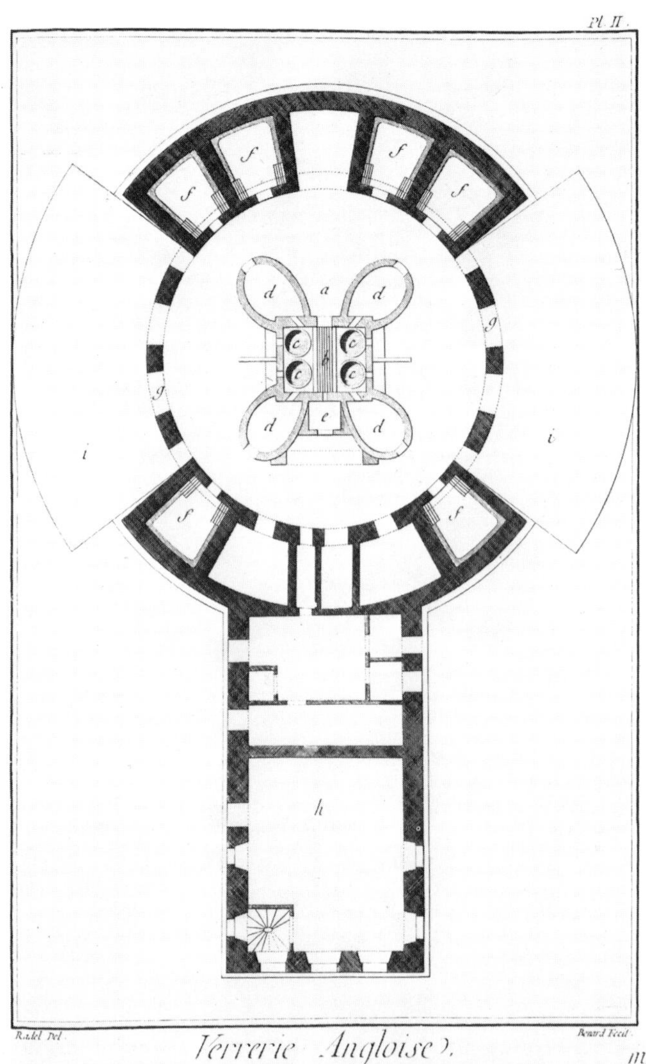

Fig 22 Plan of the ground floor of an English cone glasshouse showing the central four-pot furnace with an iron grill (b) for burning coal fuel between the sieges. (c) indicates the crucibles, and (d) the pot-arches. Four wing furnaces were used for firing pots before they were placed in the main furnace, and a small fritting oven (e) is also attached to the main furnace. Six annealing furnaces (f) flank the walls of the cone. Also visible are the linnet holes for conducting heat from the main to the wing furnaces. From Diderot and D'Alembert's encyclopaedia.

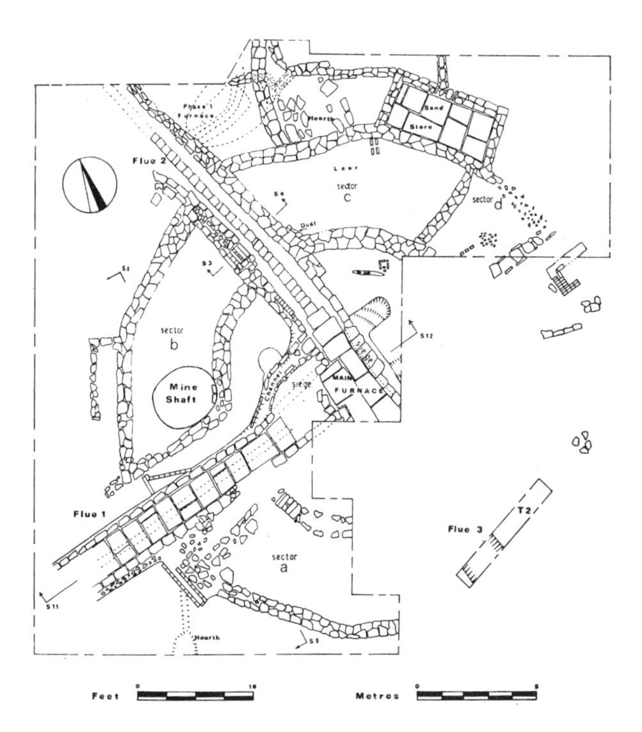

Fig 23 Plan of the cone glasshouse at Gawber, South Yorkshire, showing three flues or air intakes, feeding air from the exterior to the central furnace. S1–S12 refer to sections drawn during excavations. Site of the earlier (Phase 1) furnace is shown in the top left-hand area. T2 (Trench 2) confirmed that Flue 2 continued into Flue 3. Flue 1 was added at a later date. Sector 'a' seemed to be a smith's hearth for maintenance of glassworkers' equipment. Sector 'b' contained a mine shaft which antedates the main cone structure. Sector 'c' was probably the 'lehr' or annealing area, with a sand store adjoining the exterior wall.

An excavation of an eighteenth/nineteenth-century glasshouse at Gawber, near Barnsley, South Yorkshire, by Denis Ashurst between 1964 and 1972, also corroborated the accuracy of the French encyclopedia. Phase II of the furnace, built during a reconstruction in the 1730s, comprised a brick cone on a stone foundation surrounding a central hearth fed by three air intakes, which, with the associated lehr and raw material store, formed the major part of the works. The furnace was coal-fired and used open pots, producing phials and wine bottles in clear, green and dark brown metal, often decorated with seals (*fig 23*).[12]

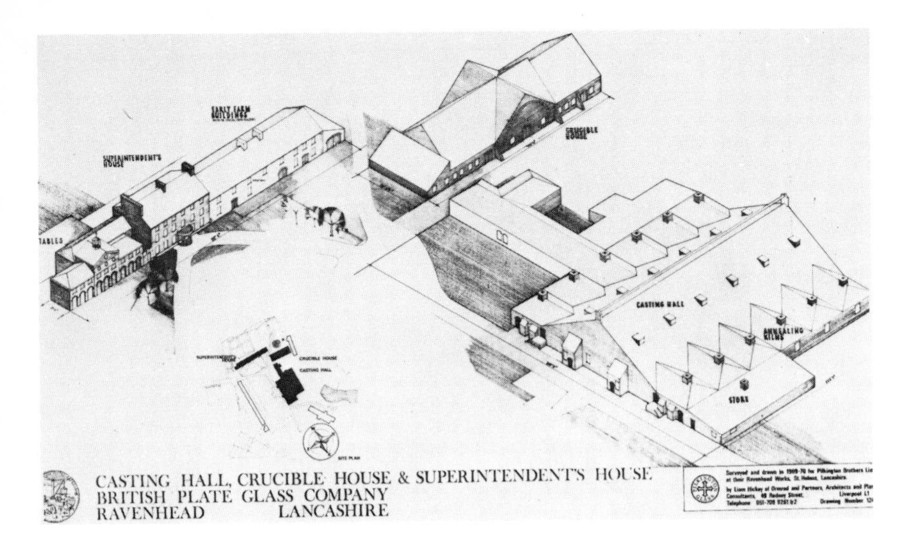

Fig 24 The British Cast Plate Glass Manufacturers' Works at Ravenhead, Merseyside, *ca* 1790, showing the casting hall, crucible house and superintendent's house.

Although this book is not primarily concerned with flat glass manufacture, mention should be made of the great Ravenhead Casting Hall at St Helens, Merseyside, which has only recently been completely demolished by Pilkington (*fig 24*). Its foundations were laid in 1773 by the British Cast Plate Glass Manufacturers, and its cathedral-like archwork was one of the wonders of the Industrial Revolution. In the 1960s and early '70s Pilkington conducted a unique survey of industrial glassmaking remains in St Helens. Careful architectural drawings were made of historic buildings like the Casting Hall, and are now kept in the Pilkington Group Archives.

1. *May 1904*
2. *Wood 1965*
3. *Wood,* forthcoming
4. *Scoville 1950*, 37
5. *Crossley 1967*, 45–83
6. *Crossley and Aberg 1972*
7. *Daniels 1950*
8. *Hogan 1968; Vose 1971, 1972*
9. *Charleston 1978*, Fig 18
10. *Westropp 1920*, 37–38
11. *Bowie 1974*
12. *Ashurst 1970*

CHAPTER SEVEN

Glasshouse Excavation

Archaeological techniques are essentially the same for any site, be it prehistoric or Victorian, a burial mound or a deserted village: but apart from the general mechanics of excavation, a certain amount of specialized knowledge is indispensable, especially on industrial sites. Besides knowing something of the history of glassmaking, a glass archaeologist needs to be aware of how the craft was carried out, what problems ancient glassmakers were likely to encounter, and how modern scientific methods can help him to unravel the secrets of old glassmaking sites.

The location of glassmaking sites

Although a certain amount of luck is attached to finding an old glassmaking site, there are a few general rules which will help aspiring glass archaeologists. There is always a greater chance of finding sites in traditional glassmaking areas of the past, such as the Weald, Stourbridge and Staffordshire. (However, sites have recently turned up in quite unlikely places, such as the Hutton and Rosedale sites on the open moorland of Yorkshire, and the Bickerstaffe glasshouse on the south-west Lancashire plain.)

G. H. Kenyon gives some first-class guidance on finding glassmaking sites. Of the forty-two proved, probable and possible sites he lists in *The Glass Industry of the Weald* (1967), half were found as a result of following up a field name. The author found the Bickerstaffe site through perusing local maps and finding a 'Glass Hey Field' in the 1841 Tithe Commutation and Award Map. Kenyon warns that one has to watch for corruptions of the word glass such as Glassus, Glasses and even Glassets.

Parish registers are another valuable source for clues to the location of glasshouses. Besides the addition of 'glassman' or 'glassmaker' with their variant spellings after names in the registers, the names themselves may also be a clue, if they belong to an old glassmaking family. Wrong spelling of a name should not act as a deterrent; for instance, the name of the Du Houx family of glassmakers was anglicized to Dehowe, Dehoe, Dehuse, De Hooke, de Hugh and de Hoc.

Before embarking on a glasshouse search in a particular area, it is worth considering what raw materials were readily available to the glassmaker to attract him to a particular locality. Many early glasshouses were built near sources of water, and near woodland that could provide fuel. Old maps may indicate where woodland once was: a study of the geology of the area will show where sand and perhaps fireclay were available. A talk with local farmers and landowners can also be very helpful to indicate a likely spot: farmworkers will be able to tell you if a lot of glass and glazed material or even burnt earth constantly appears on their fields during ploughing and cultivation. Aerial photography may also be helpful in showing soil or crop marks in fields.

Fieldwork is the most valuable way of actually locating the site – walking over a likely area looking carefully for glass and glazed fragments. Even though nothing may be seen on the first walk, it is worth repeating the exercise two or three times, especially after a hard frost which makes objects on the ground more apparent. It was at the third attempt, after a hard frost, that the first piece of glazed crucible was spotted on Glass Hey Field at Bickerstaffe.

A familiarity with the materials one is looking for is an obvious advantage, and any opportunity to examine old glass and glasshouse debris, whether at a museum or an excavation, should always be taken. Glazed stones and slag are never conclusive evidence for a glass site, since similar vitreous slags can come from iron or brick works or other industrial sites, and a few broken glass fragments can also be misleading, since broken glass, pottery and other rubbish was spread over fields with manure from the streets of towns and cities up to the advent of the internal combustion engine in this century. On the other hand, a few pieces of crucible are virtually certain evidence of glass manufacture within the vicinity of their discovery, since there is nothing else like them in any other industry. Frit or scum, often similar to slag in appearance, may also occur. Kenyon maintains that a field name indicates a glass site within about one hundred yards, a few fragments of crucible are evidence that the site is within half a mile and a concentration of glass and crucible fragments narrows the search down to perhaps twenty yards.[1]

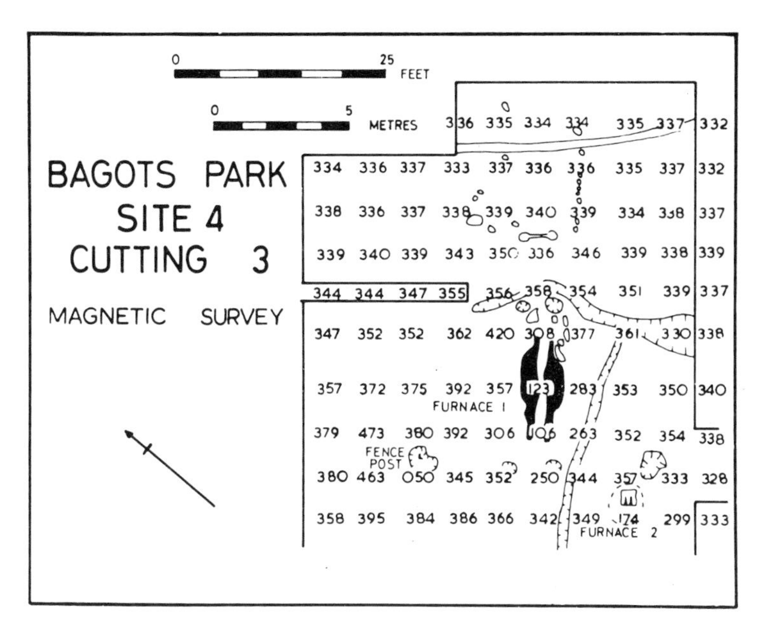

Fig 25 Results of the proton magnetometer survey conducted by Dr Patrick Strange at Bagot's Park, Staffordshire. The elongated shape of the main furnace which was to be excavated was well illustrated by three readings in line over it.

A woodland area is obviously more likely to produce the best remains of a glass furnace, despite root disturbance, since a cultivated field will have had all large stones and obstacles to ploughing removed, leaving only the ground below plough level undisturbed. A small mound in a copse where other glassmaking evidence has been found may cover the lower part and foundations of the glass furnace, and the larger the mound, the more worthwhile the excavation is likely to be.

Even when a glassmaking area has been located through a field name or scattered glasshouse debris, the prospect of trying to locate the furnace area with trial trenches can be daunting. At this point it can be very useful to have the area surveyed with a proton magnetometer. Burnt areas such as hearths or furnaces, ovens, and even buried pits and ditches, distort the normal magnetism of the earth very slightly, in the same way that iron and steel can distort a compass reading, giving a false direction for magnetic north. A compass would never register the extremely slight magnetic distortion caused by furnace remains, but a proton magnetometer does. It is capable of picking out the slight increase in the weak magnetism of iron oxide in soils and clays, caused either by heat or cultivation.[2]

Patrick Strange of the Department of Electrical Engineering of the University of Nottingham has used a proton magnetometer on several glasshouse sites. At Bagot's Park he used a detector height of fifteen inches (0.38 m), and covered three fifty-foot (15.24 m) squares, taking readings at five-foot (1.52 m) intervals (*fig 25*). A similar plan was followed at Bickerstaffe, where the instrument successfully pin-pointed the area of the furnace on the ploughed field.

Although the proton magnetometer is a great step forward in the detection of archaeological sites, it has its drawbacks, since it will register all disturbances in the ground, and not just the archaeologically interesting ones. At Bagot's Park, four anomalies in the readings turned out to be the remains of iron fencing posts. One of the most promising readings in the very disturbed mining area of Haughton Green, Denton, was found to be a buried bicycle! The more highly disturbed an area is, the less useful a proton magnetometer can be, since too many 'false' anomalies will be recorded.

A variant on the proton magnetometer is the 'bleeper', which emits a steady note over undisturbed ground, but begins to bleep over any disturbance, the bleeping becoming faster as the disturbance increases.

If it is impossible to get hold of such sophisticated equipment, copper 'divining' rods can be used as a more primitive substitute. The author can offer no explanation of why they work, but they did accurately locate the Haughton Green furnace. Two copper rods were bent at right angles, making short handles, and these were held in either hand, the longer ends at right angles to the body, and parallel to each other. Again working on a grid, the points where the rods crossed each other were recorded and the areas excavated. Personal magnetism can affect the readings: some people are always unsuccessful, and others have a high degree of success.

Resistivity surveying was used to locate the base of the cone structure at the Gawber glasshouse in South Yorkshire. Denis Ashurst, who directed the excavation, commented that the field designated as Site C was surveyed by a resistivity meter simply to locate any anomalies, and not to produce a complete picture. It was by no means certain that this was even the correct field, as the two existing eighteenth-century maps contradicted one another. Five runs of one hundred feet (30.48 m) at three-foot (0.91 m) intervals gave a clear indication of the disturbance left by the (assumed) cone structure. A five-foot (1.52 m) trench along an estimated diameter of the disturbed area confirmed that the building could probably be contained in an eighty-foot (24.38 m) square, and a full excavation could then be planned and carried out. Resistivity surveying relies on the ability of contrasting soils to conduct electricity at slightly varying rates. Water, with its mineral salts contained as moisture in the soil, acts as the electrical conductor. Differences in moisture-level make some soils better conductors than others, and the differences can be measured on a suitable instrument as variations in resistance. Soil filling a ditch will have a slightly different reading from the surrounding ground, hence changes in resistance measured over an archaeological site may be interpreted in terms of the features which caused them.

Problems in glasshouse excavation

Although the excavation of a glasshouse site is ruled by exactly the same criteria as any other archaeological undertaking, there are certain special factors which are worth taking into consideration before embarking on a glass 'dig'.

First, it is most important that the whole area surrounding the glasshouse should be thoroughly excavated. Early excavations tended to concentrate just on the main furnace and missed a great deal of the important information that can be derived from the subsidiary structures.

Since glass is fragile, two finds bowls or trays should be given to each digger, one for glass and one for less fragile finds. To avoid cuts from broken glass, safety gloves are a worthwhile precaution during excavation, and a fully equipped First Aid box on site is essential.

According to Denis Ashurst, the major difficulty encountered especially in a post-medieval glasshouse excavation is the cullet. The origins of cullet on a site must always be in doubt, and must necessarily throw suspicion on all glass fragments found there. Ashurst points out that an archaeologist excavating a pottery site tends to assume that 'wasters' and spoil heap finds are native to the site where they are found, thus providing evidence of the typology of the products and, possibly, the sequence of designs.

> This reasoning is based on the logic that broken/faulty pots are to be considered useless, and no one is likely to have brought a load from another source on to the site. Such a bias in thinking must be avoided at all costs on a glasshouse site and each piece of glass treated on its merits, with any conclusions it may suggest being ringed with caution – a glasshouse cannot be regarded as a variation on a pottery site.

Any statistical conclusions regarding types of glass found on site are of minimal value, since, unlike pottery, glass is frequently recycled. Besides the cullet produced by normal wastage at the glasshouse, the owner of a well-run establishment would never have enough waste glass for facilitating the next melt, and would be obliged to purchase old glass from any source. This could be glass of any age, from any country of origin, damaged in use and thrown out, or even parts of a waste heap from a defunct works.* The only glass that can be attributed with any certainty to a site is that found adhering to a crucible, since it is highly improbable that broken crucible or 'grog' would have been transported from another site.

With such limitations in mind, it is possible to suggest an order of probability when attempting to determine which artefacts represent the output of the glasshouse. Ashurst summarizes:

> i) Chemical analysis of glass remaining in crucibles will sufficiently identify the types of glass mix being used. Similar analysis of fragments of the glass products can be compared to this, and a range of wares identified with acceptable reliability. However if more than one glassworks were using the same formula, this could present difficulties.
> ii) If products of a similar style to these were being made, which, though displaying idiosyncrasies of finish due to a different artisan, are clearly related products, they could be accepted with a slightly lower probability, without the expense of analysis.
> iii) A range of products may be identifiable, if in sufficient quantity, by assuming that, in any collection of broken glass, a preponderance of one style is likely to represent the local product, particularly if found elsewhere on site. A purchase of glass from another works would create a misleading concentration on a cullet heap, but would be unlikely to find its way into the area used for packing up the products for despatch.
> iv) The determination of a progression of styles is dependent mainly on the quantity of glass fragments available. It might be assumed, as on a pottery site spoil heap, that the lower remains are earlier, but as the spoil heap will have been shovelled and disturbed continually in its lifetime, then the position in the heap of any fragment is no indication of its chronological position. Careful excavation of any concentration of glass fragments by small areas, keeping tally of the percentage styles, might suggest a sequence, but it is by no means certain to reward the effort it involves.[3]

Trying to build up a picture of the original shape of a glass product from fragments demands a lot of time and luck, since glass fragments

* At Knightons pieces of glass were found in the annealing furnace and in the tip of furnace waste which were thought highly unlikely to be bought-in cullet, and were accepted as being of local manufacture.

present no clues like the wheel-marks left by a potter. Many give no indication even of which way they are to be orientated. Contemporary paintings and illustrations which show glass artefacts can provide useful evidence of shapes.

The colour of a piece of glass can often help to identify it, but not by simple surface examination. The degradation of glass exposed in soil depends on composition, soil conditions and length of exposure. Some glasses deteriorate more quickly than others and the surface changes greatly affect the quality and colour of reflected light. The glass colour must be determined by looking through the glass at an even source of light. The northern sky is suggested as a ready 'standard', as varying types of artificial light seriously affect the tint observed in the glass, and parts of the same article appear as different colours if viewed through a different light source.

Fragments of coloured glass found on site, particularly in the furnace area, could well be unrepresentative; for instance, glass which has been badly fired, or accidentally subjected to secondary heating, can change colour quite dramatically and not necessarily be the intended colour. Lumps of opaque blue glass found discarded from crucibles at Gawber were the result of failure in preparing a deep amber brown.

Colourful glazes and regular 'striping', and even regular grooving, may be found on the outer side of crucible fragments, and could easily be misinterpreted as decoration. Although it is not quite beyond the bounds of possibility that a glassmaker should purposely decorate his crucibles, it would seem highly unlikely from a practical point of view – unless there was a technical reason, such as strengthening, for adding grooves to the sides. Gawber and Knightons notably produced crucibles with decoration or decorative grooving, but in my experience, every piece of crucible with stripes or grooving has been the result of the corrosiveness of hot metal dripping down the sides. It should also be remembered that crucibles and the furnace interior are never purposely glazed, but become so owing to the heat in the furnace, which causes an exudation within the refractory (*pl 46*).

Conservation on a glasshouse site

As it is impossible for an archaeologist to know, without lengthy analysis, exactly what composition glassmakers were using on any particular site, the conservation of what is virtually an unknown quantity should be approached with extreme care. Robert H. Brill, of the Corning Museum of Glass in New York, states in his informal

Plate 46 Crucible fragments from Haughton Green, showing parallel 'grooving' on the outer surface (left), and iridescent colourful 'striping' (middle and right), probably the result of hot metal dripping down the sides of the pot.

notes on field treatments that it is safest not to do any cleaning or treatment in the field if there is any doubt as to what the material is, or exactly how it will be affected. He emphasizes that the most important principle to be kept in mind by the archaeologist is that any field treatment to remove soil or accretions should not harm the glass, or remove its surface weathering.

> Probably the most important rule to follow, is that when glasses are freshly excavated, they should not be set out in the bright sun where they can dry. Certainly, on glasses coming from warmer climates, it often happens that a piece will dry out in a matter of a few minutes and lose its weathering crust almost completely. Therefore, even while pieces are in the field awaiting removal to storage, they should be kept in the shade.

John Atkinson of the North West Museum and Art Gallery Service, who organized the conservation on the Haughton Green site, says that it cannot be overstressed that the treatment of heavily degraded glass is not a job for the untrained, and irreparable damage is almost certain to result from the 'We'll see what happens' attitude, coupled with excessive impatience.

However, few archaeologists can afford the professional services of an expert conservator on site all the time, and there are certain general measures that can be taken by any careful 'digger' in the cleaning and treatment of excavated glass. Loose soil or calcareous accretions may be removed by gentle brushing with a soft brush, such as a dry camel's hair brush; but if the accretions prove too stubborn for such treatment, on no account must a knife or other sharp instrument be used, since this will almost certainly damage any weathering patina, or even the glass itself.

Although fresh water can be used quite safely to clean glass which is in good condition, it should never be used on glass which is weathered or has deteriorated. Brill observes that, for glasses which do have substantial weathering crusts, any treatment with water or other liquids or solutions may cause the crust to soften and swell and come off, either while it is wet, or some hours or days later when it has dried out. The author found the easiest solution was to avoid water completely in the cleaning of glass, since deterioration may become apparent only after water has been used and the damage has been done. Industrial methylated spirits was used to clean glass finds at the Bickerstaffe and Haughton Green sites with complete success.

Spraying or coating excavated glass with lacquers of any sort in the field is inadvisable, except in extraordinary cases. Brill states that glasses which have not weathered or deteriorated do not need it, while glasses with a weathering crust can never be restored to their original state once a coat of lacquer has been applied. Lacquers do not usually penetrate well enough to make a bond with the glass beneath, so the weathering crust is very likely to come away from the main body. Recognizing that some glasses are so very fragile that something *has* to be done to them on site, Brill says that one thing which must be avoided is the application of a thick, viscous resin, which would simply accumulate on the surface of the weathered glass, eventually removing the crust. If a consolidant must be applied, it should be dilute, very fluid and have a low surface tension so that it can better penetrate the weathering crust and help to consolidate the object as a whole. A few applications of a very dilute solution of some resin is likely to be far more effective than a single heavy application of a more consolidated solution. In Brill's view, the application of any consolidant should be regarded as an irreversible process, since the dissolving process used to 'reverse' any previous treatment is almost certain to have some effect on the weathering crust itself, either detaching it, or drastically altering its appearance.

Any unusual materials found on a glasshouse site, such as partial melts, frits, core materials, clay or limey aggregates, should not be chemically treated or cleaned with water, since they are likely to disintegrate completely, and important technical evidence may be lost.

CONSERVATION OF GLASS AFTER EXCAVATION

Storage

When polythene bags have been used as find bags on a glass excavation, it is not advisable to leave excavated material in them for very long, since any moisture left in, or on, the objects will start to 'sweat', and severe deterioration can set in. Small cotton find bags with their contents can be left safely in store until such time as they are needed. The storage area should be dry and cool. Inevitably there is some danger of breakage when a number of glass objects are 'bagged' together and have to be transported from site to storage. It may be advisable to place the better glass finds in specially prepared cardboard boxes with strong polyurethane foam filling, slit with a scalpel to enclose individual glass objects with maximum firmness, yet with no danger of breakage. This method has the added advantage of allowing excavated glass to be viewed quite safely, and easily.

As most crucible is very tough, this can be stored loose in strong cardboard boxes, though care must be taken not to put in more weight than the box can stand.

Repair and restoration

The very properties which make glass useful to man, i.e. its very dense nature and lack of porosity, also make it one of the most difficult substances to restore. The number of adhesives suitable for glass is limited, and their success can never be guaranteed in the long term. There is also the added danger with some new conservation methods of having to subject glass objects to heat, which can have disastrous results if there are any hidden stresses in the glass.

According to John Atkinson, the only adhesives which give a reasonable chance of success are the epoxy resins. He has used the five-minute colourless, the rapid and some other grades of epoxies, and a few of the cynoacrylates, but he stresses that not all of these are suitable, and the manufacturers should be consulted if there is any doubt at all. Denis Ashurst maintains that a clear epoxy resin has been found to be the only satisfactory method of restoring heavy glass with a clean break. He suggests that for thin-walled vessels, where the joining surfaces are friable, the problem can occasionally be solved by extending a smear of resin inside the glass for about 0.5 cm each side of the break to add a little strength. However, for many shattered items, a drawn reconstruction remains the only answer.

When glass is badly weathered, flakes of iridescence will fall very easily from it, causing a general weakening. Many transparent fillers such as varnishes, waxes and resins have been used on such glass for strengthening, while still preserving its transparency. According to Atkinson, the substance must be easily applied, reversible and chemically stable so it does not react with the glass; must possess a smooth hard surface; produce no marked shift in colour tone; involve no heat, or very gentle heat, and no shrinkage. Over the the past fifteen years he has used a solution of 5% PVA (polyvinyl acetate) in industrial methylated spirits or acetone, which has produced acceptable results. Several coats of the solution may be applied when the previous one has dried, and the preparation can be easily removed.

The methods of rebuilding an object in glass are similar to those used with ceramics – sand tray, sticky tape, a steady hand; and an unlimited supply of patience.

Crisselling

Crisselling in glass is often referred to as 'glass disease', starting with the formation of tiny, hair-like cracks in the metal, which can be almost undetectable, advancing to coarser cracking when the glass becomes sticky and 'sweats', and finally reaching a state where the glass becomes opaque and milky, and is so badly crisselled that flakes of glass fall off when the glass is handled. The condition is caused by a breakdown in the chemical structure of the glass due to an excess of alkali. If glass is found in this condition on site, there is little one can do except add silica gel as a dehydrating agent to its packing, and send it promptly to a conservation laboratory. No one has yet found a permanent cure for crisselling, and the most one can do is attempt to stabilize the glass so that any further deterioration is unlikely.

Organic lacquers were used to form a protective film on glass as early as 1903, but, as these lacquers have been found permeable to moisture vapour, they cannot be regarded as a permanent cure for crisselling. Brill observes that most treatments for crisselling involve impregnations with polymeric materials, which consolidate the surface and restore transparency, but none has yielded entirely satisfactory results.[4] The 'weeping glass' at the British Museum is kept in special cases, where the relative humidity is kept below 42% by means of silica gel with a small fan at the top of the case to ensure adequate circulation. Individual pieces can be kept in air-tight boxes with silica gel, where the gel will have to be changed only about once every six months, when it turns from a blue (active) colour to pink (exhausted). For glasses with incipient crisselling, the Corning Museum of Glass has designed sealed cases with 45–55% relative humidity range, with provisions to shift the humidity level in either direction – for instance, if persistent slipperiness developed on the glass surfaces, the humidity could be dropped slightly below the 42% level.

An elaborate method was used in Sweden in 1962 to conserve the Kungsholm glass from the glasshouse founded there in 1676 (see p 87). The glass had begun to 'weep'. All grease was removed from its surface with carbon tetrachloride, and then it was washed, first in a very dilute nitric acid, then in distilled water, followed by immersion in two baths of alcohol or industrial methylated spirits. After drying, the glass was put in a cylinder of Perspex, which was evacuated and filled with argon and made air-tight. Although the glass was then in a permanently inert atmosphere, it became somewhat difficult to inspect it closely.[5]

SCIENTIFIC DATING

Analytical methods of dating glass

It was the distinguished Swedish chemist, Martin Heinrich Klaproth, who first described the use of analytical methods in archaeology, with a lecture given to the Royal Academy of Sciences and Belles-Lettres in Berlin on 4 October 1798. His subject was a report on the results he had obtained in the analysis of samples of ancient glass. Analytical research continued somewhat spasmodically during the nineteenth century, but it was not until the development of refined methods of analysis on a micro scale in this century that intensive application of analytical methods in archaeology became possible.

The purpose of analysis of glass and other materials is twofold. First, systematic analysis involving large numbers of samples may provide statistical information about the original sources for raw materials, information on old trade routes for the raw materials, changes in ancient production methods, or the relative dating of some classes of artefacts. Secondly, specific analysis of a single object or small group of objects may help solve special problems relative to the objects, including questions of authenticity. However, unless a problem has been narrowed down in such a way as to make investigation reasonably decisive, scientific analyses should be used sparingly, owing to their high cost. There is not much point in making analyses unless there is an existing background of analysed material for comparison.

Systematic analysis has been used effectively to obtain significant information on glass. Between the time of Klaproth's pioneering work and 1968, over 300 samples of ancient glass from Egypt, the Far and Near East, the Roman Empire, India, medieval and Renaissance Europe and Russia have been analysed, and have shown that different basic formulations were often used for making glass in different periods and regions, and that special ingredients were added to create colour, transparency and other attributes of glass.[6] The analytical results obtained can serve as a basis to assist an archaeologist in differentiating between glasses of different periods and regions.

Modern analytical techniques which have been used successfully with ancient glass include X-ray diffraction analysis, which is particularly useful for identifying the materials responsible for producing opacity or translucency in glasses; isotopic analysis, which can be used on any lead-containing artefacts, including glass; and electron beam microanalysis, which will be described in more detail later in the chapter.

Specific analysis carried out on the famous dichroic Lycurgus Cup in the British Museum revealed that minute quantities of silver and gold within the glass material were primarily responsible for the Roman glass having a wine-red colour by transmitted light and an opaque green colour by reflected light. The microchemical techniques used to analyse the glass included spectrophotometry, flame emission photometry, and X-ray fluorescence spectrometry, which achieved a complete quantitative analysis on a total sample of less than nine milligrams. A double achievement was that Brill was enabled to reproduce the dichroic effect of the Lycurgus Cup in a glass bowl, thus re-creating the ancient technique.[7]

Detection of forgeries can be another function of analysis of glass. An alleged Roman glass which had a 'weathered' appearance was found by X-ray diffraction analysis to have artifically induced weathering, brought about by treating the glass with hydrofluoric acid. The same technique revealed that the opacifying agent used in the body of the glass did not correspond with the type of opacifying agent which has been shown to be characteristic of glass from the Roman period.[8]

Spectrographic analysis

This is one of the most widely used analytical techniques and is rapid, relatively unambiguous and non-destructive. This can be of particular importance to an archaeologist wishing to analyse rare specimens, and some of the instrumentation is sufficiently portable to be used on site. Spectrographic instrumentation is very expensive, but can be found with most steel companies, some glass companies, universities such as Oxford, Cambridge, Birmingham and Loughborough, and large museum laboratories, such as the British Museum Research Laboratory.

Spectographic analysis relies on the fact that, under normal circumstances, the electrons surrounding any atom are in stable orbits, but, if exposed to some energetic stimulus such as heating or bombardment with energetic radiation, the electrons can be disturbed and may be lifted into higher orbits, or orbits of greater radius. An atom in such a state is unstable and inevitably the electron falls back to its original orbit, emitting energy of slightly longer wavelength than the stimulating source. The energy or radiation emitted by atoms returning to their normal state is characteristic of the atom in question, and, if analysed, can be used to identify the type and quantity of atoms present in any chemical element. There are various types of spectrographic analysis, each type depending on which electrons have been disturbed and what exciting radiation has been used, e.g. optical spectra, infra red spectra, X-ray spectra and γ-ray spectra. As with other microanalytic techniques, the basic importance of spectrographic analysis is to determine the constituents of a sample, which can then be dated by comparison with analyses of samples with known dates. It is often the presence of minute traces of certain elements, rather than the main constituents, that characterize location and date.

Electron beam microanalysis

The development of the electron microbeam probe began about 1948 and has proved a powerful technique for performing microchemical analyses. Its application to glass analyses was pioneered by Brill and Sheldon Moll, who first published their successful researches in 1961.[9]

The electron microbeam probe focuses a beam of high energy electrons on to a tiny spot of the sample. The atoms at, or near, the surface of the glass are excited to higher energy states and emit, or fluoresce, X-rays. The X-ray spectrometer unit collects these X-rays and performs two measurements. By separating the X-rays into their component wavelengths, it records which wavelengths have been fluoresced and can thus determine what chemical elements are present in the sample. (This is because atoms can emit X-rays of only precisely defined wavelengths, which are unique to each element.) The instrument also measures the intensity of each wavelength, which evaluates the amount of each element in the sample.

The chief advantage of the method is that it is virtually non-destructive, the only necessary preparation being that the surface of the sample has to be polished with a metallographic polishing wheel. The instrument can be focused on an area as small as 1000 to 2000 nm in diameter (0.001 – 0.002 mm), which means that it can be used to analyse objects which are too small to be seen with the naked eye. Brill and Moll used it for quantitative analysis of small regions of glass objects, identification of opacifiers, analysis of stones and metallic inclusions in glass, analysis of weathering crusts and measurement of concentration gradients in ancient glass.

Surface analysis

The last few years have seen the emergence of other spectrographic techniques using the same basic principles – the use of an exciting beam of energy, and the detection and analysis of radiation or particles excited in the surface of the specimen. These newer methods, which include Auger Electron Spectroscopy (AES), Electron Spectroscopy for Chemical Analysis (ESCA), and Secondary Ion Mass Spectroscopy (SIMS), are more accurately termed techniques of surface analysis in that the depth of the excited region in the specimen is much smaller than with, say, electron beam microanalysis. They are therefore much more sensitive methods of depth profiling. In addition, AES and ESCA can be used to give information on the type of chemical bonding within the excited region.

The following table* indicates the relative difference between these techniques.

Fission-track dating

Fission-track dating has been used for both geological and archaeological dating from the early 1960s, and the possibility of its use for dating man-made glasses was first reported in 1964 by Brill of Corning, New York, and Drs R. L. Fleischer, P. B. Price and R. M. Walker of the General Electric Research Laboratory, New York.[10] The technique relies on the fact that most naturally occurring materials, including natural glasses such as obsidians, tektites and impact glasses, contain very minute quantities of uranium, which is fissionable (capable of

* Reproduced by courtesy of Gerald Shaw, Principal Technologist, Pilkington Brothers Ltd.

	AES	ESCA	SIMS
Exciting energy	2 to 5 kilo electron volts	soft X-rays	ions of a few kilo electron volts
Detected emission from the surface	characteristic secondary electrons	characteristic secondary electrons	sputtered secondary ions
Lateral resolution	down to 50 nanometres	greater than or equal to 1 mm	greater than or equal to 1 mm
Depth resolution	surface (c. 0.5 nm)	0.5 to 3.0 nm	surface (c. 0.2 nm)
Ease of quantification	Easy	Easy	Very difficult
Elements detected	Any element other than hydrogen or helium	Any element other than hydrogen or helium	All elements
Usage	fairly common in universities and industry	fairly common in universities and industry	in infancy
Chemical state information	differences in valency state easily detected but not for all elements	differences in valency state readily detectable for almost all elements	can give information on molecular groupings; good for adsorbed organics
Limitations	some beam damage on insulating specimens e.g. glass and ceramics; erratic emission from rough areas	difficult to identify and locate selected areas on specimen; peak broadening occurs with insulating specimens; depth profiling very slow	beam positioning problems as in ESCA; unstable emission from insulating specimens; qualitative information only, very slow
Advantages	very good for depth profiling and selected area work; very fast unambiguous technique	unambiguous technique	useful for identifying adsorbates and in catalysis work

nuclear fission). During the long history of these materials, the naturally occurring uranium, which consists in the main of an isotope U_{238}, spontaneously breaks up into fragments, accompanied by a simultaneous emission of neutrons. The fragments have some kinetic energy and break through the vitreous lattice in glass, until this energy is expended. This decay has no visible effect on the glass, but if a freshly fractured surface, or a polished surface, is given a mild etching treatment with hydrofluoric acid, the damage, or microscopic etch pits where the uranium atoms have disintegrated on the surface, will show up as 'fission tracks'. If the etch pits are then counted microscopically, and the uranium content of the glass is known, it is possible to use the known rate of decay for uranium to calculate the age of the glass, or, more correctly, the time that has elapsed since the glass was last heated. (Heating the glass will anneal or repair the undeveloped fission-damage centres.)

Instead of removing a fragment of glass for chemical analysis to find out its uranium content, the researchers found it simpler to put the same sample from which they had taken the etch pit count into a nuclear reactor and bombard it with a known dose of thermal neutrons. This causes more 'fission tracks' to appear, since the other isotope of uranium which is present (U_{235}), does not split spontaneously, but only under the action of slow neutrons, and it can then be determined what concentration of U_{235} atoms is present. This is induced fission, contrasting with the spontaneous fission described before. Hydrofluoric acid is again applied, and the increased number of etch pits counted, from which it is possible to calculate the age of the glass.

Scarcely any damage to a glass object is entailed, since only a light, transient radioactivity is introduced into samples, and a chip of glass a few millimetres on each edge, or even a small patch, polished and etched on an inconspicuous surface of an object, is all that is necessary for the technique to be used.

The usefulness of fission-track dating diminishes to almost nothing for ancient man-made glasses, since their uranium content falls below the required level for the technique to work, and the rate of spontaneous decay of uranium is very slow. Natural glasses contain about one part per million of uranium, which, in effect, means that one atom of uranium can be expected to disintegrate every two years in each cubic centimetre of these glasses. The great age of these glasses means that the number of etch pits is sufficient for dating to be possible, but man-made glasses are less than a thousandth of the age of most natural glasses and have to contain considerably more uranium to be suitable for fission-track dating. As far as ancient glasses are concerned, the only possible use for this dating technique is for verifying the antiquity of questionable objects, since only very rough age estimates are feasible.

Fission-track dating comes back into its own as an accurate technique in the more modern period of glassmaking, when glassmakers began to add uranium to their products as a colorant. In 1789 Klaproth first recognized uranium as a chemical element and is reputed to have intentionally added it to glasses as a colorant. It was to be another half century before the use of uranium in glassmaking gathered momentum, and it has frequently been used since then, usually producing a bright, yellow-green glass, which fluoresces under ultraviolet light, or a brownish or amber glass. Brill, Fleischer, Price and Walker used fission-track dating on examples of these later glasses, which was in fair agreement with the known dates of manufacture.

A modern development has been to apply the technique to vitreous inclusions in pottery, such as particles of obsidian, zircon, mica and glazes, which have been found to give a fair correlation with known dates for the pottery in question.[11]

Weathering crusts

From as early as 1863 scientists have been aware that the weathering crust found on some ancient glasses has a stratified structure, but it was only in 1961 that the possible potential of this phenomenon for dating glass was fully realized by Brill and Harrison P. Hood at Corning, New York. Their work indicated that glass objects which have been buried or submerged for some time can undergo a chemical deterioration, producing a crust of seasonal rings, which, when counted in section under a microscope, can give the precise date of burial. At first it seemed that a dating system for glass as reliable as dendrochronology for dating wood (counting annual growth rings in ancient trees) had been found, but subsequent research, notably by R. G. Newton of the British Glass Industry Research Association, Sheffield, has cast doubt on the reliability of the method.

Weathering crusts can be observed in most collections of ancient glass. They can differ in colour, texture, mechanical strength; they may be glassy, dull, hard, soft, adhere tightly to the body of the glass, or be so fragile that they will fall off at the lightest touch. The deterioration starts at the surface of the glass, so that the outer surface of the weathering crust coincides with the original outer surface of the glass. In effect, this means that the earliest layer is the outer one, in contrast to tree rings, where the innermost ring counts as the earliest date. Although the method is destructive, it calls for only small samples, and the poorest fragment is as valuable as a complete object for dating. It should be emphasized that the technique does not give the date of manufacture for the glass, but the date that the glass object was first buried or subjected to weathering.

It is impossible to see the layers in a weathering crust, even with a good hand lens, since they measure about 1/30 the thickness of a human hair. Only using microscopic examination or a photomicrograph of the sample can the painstaking task of counting the layers be undertaken (*pl 47*). Theoretically this should give an accurate dating

Plate 47 Photomicrograph of a section through the weathered crust of a linen smoother found at Old Erringham, showing clearly the weathering layers which dated the glass to between 1220 and 1300 AD, in agreement with associated pottery.

for the burial of the glass ± twenty years. Using weathering crusts from glass samples which had been buried or submerged for known periods of time, Brill, Hood, Newton and Shaw have all come up with amazingly accurate datings. For example, a Dutch gin bottle recovered from a well in Jamestown, Virginia, dated 1600–1640 from its style and objects found with it, revealed a date of 1646 plus or minus ten years from its weathering crust; and a wine bottle fragment from

Port Royal, Jamaica, submerged in the sea by earthquake in 1692, revealed a date between 1685 to 1701 by this technique, with a most probable date of 1691.[12]

It is not known what causes the layers in weathering crusts, although it was originally thought that they were the result of either the variation between summer and winter temperatures, or the alternation between dry and rainy seasons. Where glass has been submerged in water or buried in localities where there is relatively uniform rainfall, the change of temperature from summer to winter seemed sufficient to cause layering. However, Lucy E. Weier in her paper 'The Deterioration of Inorganic Materials under the Sea' has put forward another theory. She said that the glass found submerged in sea water at Port Royal, Jamaica was probably lying in a reducing sediment, where there was a good possibility of bacterial action.

Bacterial action can have a seasonal effect on the pH of a sediment, conditions being more acidic in the autumn and winter because of the decay of organic matter. There is a very definite silicon cycle in the sea where growing phytoplankton populations extract vast quantities of silicon from the water in the spring, with a short second spurt sometimes occuring in September, after which the organisms die and fall to the sea bed, replenishing the silicon supply. Weier goes on to ask if such factors might affect the diffusion of alkali and silicate ions in glass buried in the sediment, silicate ions being affected in the spring, and alkali ions being affected in the autumn, thus resulting in the formation of 'weathering' crusts.[13] Her theory seems more convincing than Brill's suggestion that temperature is the controlling factor, and certainly bears further investigation.

Newton pointed out that glass from the Bagot's Park glasshouse site in Staffordshire could be about equally divided into glass with heavy weathering and glass with unaltered appearance, all coming from the same archaeological context, and all more or less of the same chemical composition. From this, he concluded that the weathering phenomena in glass were so complicated that an examination of the crusts and weathering layers would not be useful for dating glass. However Weier has suggested that localized bacterial activity in the micro-environment may explain the apparent paradox of the Bagot's Park glass.

Newton observed that accelerated growth of weathering layers has been achieved under unvarying conditions in an autoclave. Ten weathering layers were produced on a high alkali glass which was exposed to the action of CO_2-free steam in an autoclave for four hours at four atmospheres pressure (144°C).[14] However Gerald Shaw, under the most extreme conditions of microscopic examination, counted five weathering layers on a piece of modern glass which had been buried for five years.

Much controversy has also raged around some layers which are hardly uniform, and resemble something akin to geological unconformities. The so-called 'plugs' of weathering which intrude through the regular layers often yield different dates, which again puts the accuracy of the technique in doubt.

Many glass samples are not suitable for dating by this method. First, the weathering layer must have been preserved intact; and since, up to a few years ago, the tendency was to clean excavated objects at once to reveal the base glass underneath, there is very little ancient glass that would qualify for examination for this reason alone. Some ancient glasses are resistant to corrosion and have produced no weathering crust, and others are so fragile that it would be dangerous to attempt the technique. The presence of enamelling on glass can also interfere with the development of weathering layers.

Dating by weathering crusts certainly cannot be dismissed simply because no one has yet been able to explain why it can, and does, work in so many cases. That inexplicable anomalies exist within this dating system is a fact of which everyone should be aware, and be wary whenever it is used. It could be used for supplementary dating evidence for glass and glasshouse sites, but never as the only dating factor. The experimental procedures involved are quick, straightforward and comparatively cheap, and it is to be hoped that future research will remove the uncertainty surrounding this dating method, which in 1961 promised to be such a breakthrough for glass archaeology.

Radiation dating

A method of identifying medieval potash glass in situ, by the detection of its natural radioactivity, has been developed recently by A. P. Hudson and R. Newton. The main experiment was carried out successfully on windows in York Minster, where background radiation was low. Experiments with standard radiation monitoring films showed that an exposure period of two months gave promising results. Glasses from the twelfth to sixteenth centuries gave count rates between three and four counts per second (cps); seventeenth and eighteenth-century glasses gave rates between one and two cps, and later glasses gave no distinguishable count rate. Experiments with a waterproof pack incorporating the measuring film and a control film for background radiation showed that fading caused by water and rain was not excessive. Hudson and Newton claim that the method is a cheap, convenient and reliable means of discriminating between high and low potash glasses, using periods of exposure of three months.

Dating by glass hydration

A new non-destructive method for dating or authenticating man-made glass objects has been under investigation by William A. Lanford of Yale University. From the time that glass is made, it adsorbs water into its surface, but so slowly that even after a thousand years the water layer may be only a few micrometres thick. This method is based upon the relation between the age of a glass object and the thickness of the layer of hydrated glass on its surface. Lanford has found that nuclear resonance, a technique which measures the hydrogen atoms of the water at various depths from the glass surface, shows a good correlation between specimens of known age, and the degree of water penetration. His preliminary studies have demonstrated that the method can help to authenticate glass objects. Initial results were based on profiles of rather recently produced glass, but it would seem that the method could also be applied to ancient glasses, although surface corrosion present on excavated glasses could make reliable hydration dating difficult. Drawbacks are that the equipment needed for hydration dating is extremely expensive, and considerable expertise is necessary in its operation.[15]

Archaeomagnetic dating

Archaeomagnetic dating relies on the fact that the magnetic field of the earth is constantly changing in direction and intensity. It has been discovered that clay which has been subjected to heat retains the magnetic field prevailing at the time of its last firing. If burnt clay is found in an undisturbed state on a glasshouse site, it should be possible to compare and measure the deviation, or 'angle of dip', of the ancient magnetic field with the modern one, and thus give it an approximate date. M. J. Aitken and H. N. Hawley of the Research Laboratory for Archaeology, University of Oxford, have been responsible for the magnetic dating of later glasshouse sites, and have provided a new and exciting dimension in archaeological dating which is by no means confined to glass archaeology.

Thermo-remanent magnetic dating, as it is also known, gave a *ca* 1330 date for the Blunden's Wood glasshouse site, which conclusively established it as the earliest excavated glasshouse site in Britain. Knightons, Alford, glasshouse was dated around 1550, with the additional evidence of magnetic dating. Bagot's Park glasshouse received a confident magnetic dating of 1535 ± 35, and the final firings of Hutton and Rosedale glasshouse sites were assessed to be of approximately similar date, probably in the last quarter of the sixteenth century, although it was not possible to say which was the earlier of the two, or whether they were exactly contemporary. Thirty-eight orientated samples were extracted from the glass furnace at Bagot's Park and ten samples were taken from the annealing hearth to the laboratory at Oxford, which provided sufficient corroborative evidence to give a confident dating. At the Gawber glasshouse, samples could be taken only from the north face of the furnace, since the south face had cracked in weathering and was considered unsuitable, but nevertheless a reasonably confident dating of 1690–1735 was achieved.

If it is decided that a glasshouse site may be suitable for magnetic dating, the archaeologist has to remember that sample-taking should begin at an early stage in the excavation, since magnetic measurements of the burnt clay or brick must be taken while they are still in their original undisturbed positions. The Haughton Green site was not considered suitable for magnetic dating, since the likely presence of an iron grill within the furnace would have upset the magnetic readings, combined with the disturbed nature of the site, which had been the scene of subsequent coal mining and the laying of at least two sewer lines.

THE RECONSTRUCTION OF GLASSHOUSES

An archaeological report on an excavated glasshouse, though no doubt excellent in itself, does not have much visual impact as far as the non-archaeologist is concerned. Of necessity, and in common with all archaeological reports, it is a record of the steady dismantling of the structure it seeks to describe. No doubt in reaction to this, a few rare attempts have been made to piece together excavated glass furnaces, to give a three-dimensional view of what they were like. Although the idea of attempting to reconstruct a glasshouse from excavated remains is a good one, there is a certain amount of unhappiness amongst purists when conjectured reconstruction becomes intermixed with, and sometimes takes over from, excavated evidence.

The happiest compromise is to rebuild the glasshouse exactly as it was found, minus covering debris, and to have by its side a model or drawings of its conjectured appearance when fully operational. The original stone or brickwork is there for everyone to see, and the conjectural reconstruction model can be accepted, dismissed or improved on, without interfering with site evidence.

The Bishop's Wood furnace in Eccleshall, Staffordshire, is a gem amongst excavated glasshouses, since it is the best survival of a forest furnace in the British Isles. Virtually no restoration was carried out on it, but only preservation, according to its discoverer, T. Pape; although some doubt is now felt about its complete accuracy.

The Jamestown Glasshouse in the Colonial National Historical Park, Jamestown, Virginia, U.S.A., is by far the most spectacular reconstruction of a forest glasshouse ever attempted. The excavation of the glasshouse revealed the remains of one main furnace and three subsidiary ones, two of which appeared to be joined. The function of the three smaller kilns was not clear, but J. C. Harrington suggested that the independent small furnace was a pot furnace, and the two adjoining ones were for fritting/annealing and annealing respectively.

Plate 48 Rosedale glasshouse reconstruction at the Ryedale Folk Museum, Hutton-le-Hole, North Yorkshire, showing the main furnace, looking between two of the auxiliary 'wing' furnaces. A synthetic 'clay' dome has been added over part of the central furnace and one auxiliary furnace.

His interpretation has been faithfully followed in the reconstruction at Glass House Point, about fifty yards from the original site.

Boulders for the reconstructed furnaces were brought from upstream along the James river, a source possibly used by the original glassmakers in the early 1600s. Reeds from the nearby marshes were used to thatch the roof of the 'cruck' building, sheltering the furnaces, just as the colonists would have done. An abundance of large trees was available to the glassmakers, and the Jamestown Glasshouse Foundation in 1956 used suitable curved tree trunks to build the shelter over the glasshouse. Although there was no archaeological evidence for the thatched timber shelter with open sides, Sidney King's 1622 painting of Jamestown depicts such a building over the glass furnaces, with a central aperture in the roof, through which smoke is rising. Sadly the Jamestown glasshouse reconstruction was burnt to the ground in October 1974. It was rebuilt, this time with a tiled roof, in May 1976.

The first British attempt to reconstruct and display an excavated glass furnace occurred between 1969 and 1970, when the Rosedale glass furnace was transported to the Ryedale Folk Museum, Hutton-le-Hole, North Yorkshire. The Curator, B. Frank, was responsible for moving the furnace, and for its reconstruction in the museum grounds. The project started as soon as the excavation of the four-winged furnace and auxiliary furnaces was completed in 1969. The stones of the furnace, some of which weighed half a ton or more, were moved from the site at the bottom of Scugdale, first to a more accessible spot on the top of the moor, and finally to the museum. Wherever possible, the stones were numbered and their positions recorded on measured drawings. Using these as a guide, as well as photographs taken of the furnace, Frank and his helpers managed a faithful reconstruction of the Rosedale furnace. Many of the stones which had been subjected to furnace heat were in a brittle state, and great care and many tedious hours were spent in treating the fragile stones and piecing small sections together. By August 1969, the work had progressed sufficiently for Frank to start building a wooden shelter to protect the furnace, which, though probably more substantial than any the glassmakers had, was copied from Elizabethan industrial buildings.[16]

The Rosedale glasshouse reconstruction, with accompanying displays of glass and crucible from the site, was opened to the public in May, 1970. The addition of synthetic 'clay' domes to the original structure to illustrate what the upper structure may have looked like is a little doubtful, both in terms of display and historical accuracy (*pl 48*). Nevertheless, the Rosedale reconstruction is a unique piece of work and a tribute to the enthusiasm and dedication of local volunteers and the Trustees of the Ryedale Folk Museum.

Possibly the most elaborate reconstruction project was carried out at the Haughton Green, Denton, site, near Manchester. The furnace itself was judged to measure approximately thirty-five feet (10.6 m) by eight feet (2.4 m), involving many tons of stonework. A complex system of conservation, cataloguing, recording and photography had to be worked out on site to ensure that stones would eventually be returned to their exact original position.

As the furnace had to be kept perfectly dry during the dismantling process, a polythene greenhouse, specially fastened to metal skids at its base for ease of removal, was constructed over the area.

The conservation of stones was of great importance, since many had suffered varying degrees of heat with consequent brittleness, the worst being like loose sand. PVA was sprayed very gently on to the top of these stones; they were then allowed to dry, turned and sprayed again, so that their physical strength was greatly improved. Stones which had broken were reattached with PVA solution of about 50% to 60%, as heavy as was practical. After being consolidated, the stones were then numbered in the normal way: an area on one surface was cleaned, dried and then lacquered, the appropriate numbering was applied in white paint, and a further coat of lacquer applied once the paint had dried. On no account could stones be moved when wet, since, although their weight increased, their physical strength decreased almost to zero. No stone was removed until its position had been drawn, photographed in both black and white and colour, and catalogued.

A major problem in the dismantling of the dry stone walling of the furnace came with the central flue arch: there was a danger that as soon as the keystone was removed, the whole structure might collapse. To avert this catastrophe, Atkinson placed a long, flat wooden board inside the flue, which closely fitted the sides and length of what remained of the structure, and was packed down where necessary. A large, heavy-duty polythene bag filled with polyurethane foam was placed in the space left beneath the flue. The foam expanded rapidly, releasing a great deal of heat and poisonous gases, and remained too hot to touch for some time. After twenty-four hours the stones of the arch could be moved with ease and safety, the foam providing a very strong base, and also providing a mould for the arch when it was reassembled.

Once the arch and the upper portions of the flue walls had been dismantled, the biggest task remained – the consolidation and removal of the remaining one and a half ton siege bank, which had to be taken

Plate 49 Conservation officers at the Haughton Green site, John Atkinson and John Price, carefully drilled underneath the one remaining siege, covered in silver foil, in readiness for the iron bars which would provide a lifting base during its removal.

away in one piece. Over a period of three weeks, a large volume of PVA solution (PVA in alcohol and acetone) was pumped into the siege through cracks and then allowed to dry. The siege was wrapped in two layers of aluminium foil, packing first being inserted over any sharp corners, and then secured by sticky tape. Holes were carefully drilled underneath the siege through its width on a predetermined line, and iron bars driven through to provide the lifting base (*pl 49*). Two wooden beams were placed under the bars, and jacks fixed at each corner.

A large plywood crate lined with polythene had been prepared to take the siege, constructed so that any one panel could be removed. The sides were screwed together around the siege and the gaps were filled with polyurethane foam. The siege was carefully jacked up, and a pallet with slings slipped underneath it. When the top lid had been bolted down the crate containing the siege was lifted by crane out of the trench and on to the side, where it was turned over, the bars withdrawn, and the bottom, which now became the top, secured. It was gently lifted once more by crane and placed in a van equipped with a hydraulic lift. An estimated fourteen tons (1640 stones) of the furnace structure was brought back to the Pilkington Glass Museum Store at Sheetworks in St Helens.

The following year, a further twenty tons of stonework was removed, including the rest of the flue structure, which had been partially built into the natural clay and rested on large foundation stones. Cobbling on the floor area of the siege was removed under difficult conditions, since the natural water table in the valley of the River Tame continually submerged the bottom of the flue, and pumps and wooden blocks had to be used to hold back the water.

The final reconstruction of the Haughton Green furnace by the Pilkington Glass Museum has yet to take place, but the cost of the dismantling operation alone ran into five figures. However, the smaller Rosedale reconstruction cost only a few hundred pounds; so it might well be feasible for an enthusiastic amateur archaeological group to reconstruct a similar forest glasshouse.

Although this book should provide a sound reference to the archaeology and history of glassmaking, new developments are going on all the time. Much is left to be discovered. The excavations described here probably make up only a small fraction of what remains to be unearthed in both town and country. New sites turn up nearly every year, and it is hoped that this book will provide the inspiration and background knowledge for the discovery of many more.

Advances in science and glassmaking technology have had a major effect on the subject: conservation and dating techniques are becoming increasingly sophisticated. The glass archaeologist has today to be an all-purpose animal – historian, scientist, draughtsman, photographer, diplomat and labourer, with a dash of the patience of a saint. Newcomers to the field can be certain that there are exciting years of discovery ahead.

1. *Kenyon 1967*, 147
2. *Aitken 1974*
3. *Ashurst,* letter to the author, 23 December 1975
4. *Brill 1975*, 121
5. *Werner 1966*, 48
6. *Werner 1968*, 34A
7. *ibid*, 38A; *Chirnside and Proffitt 1963*, 18–23
8. *Werner 1968*, 40A

9. *Brill and Moll 1961*, 260–76
10. *Brill, Fleischer, Price and Walker 1964*, 151–7
11. *Nishimura 1971*
12. *Brill 1961*, 19
13. *Weier 1973*
14. *Newton 1971*, 2
15. *Lanford 1977*
16. *Hurst 1970*

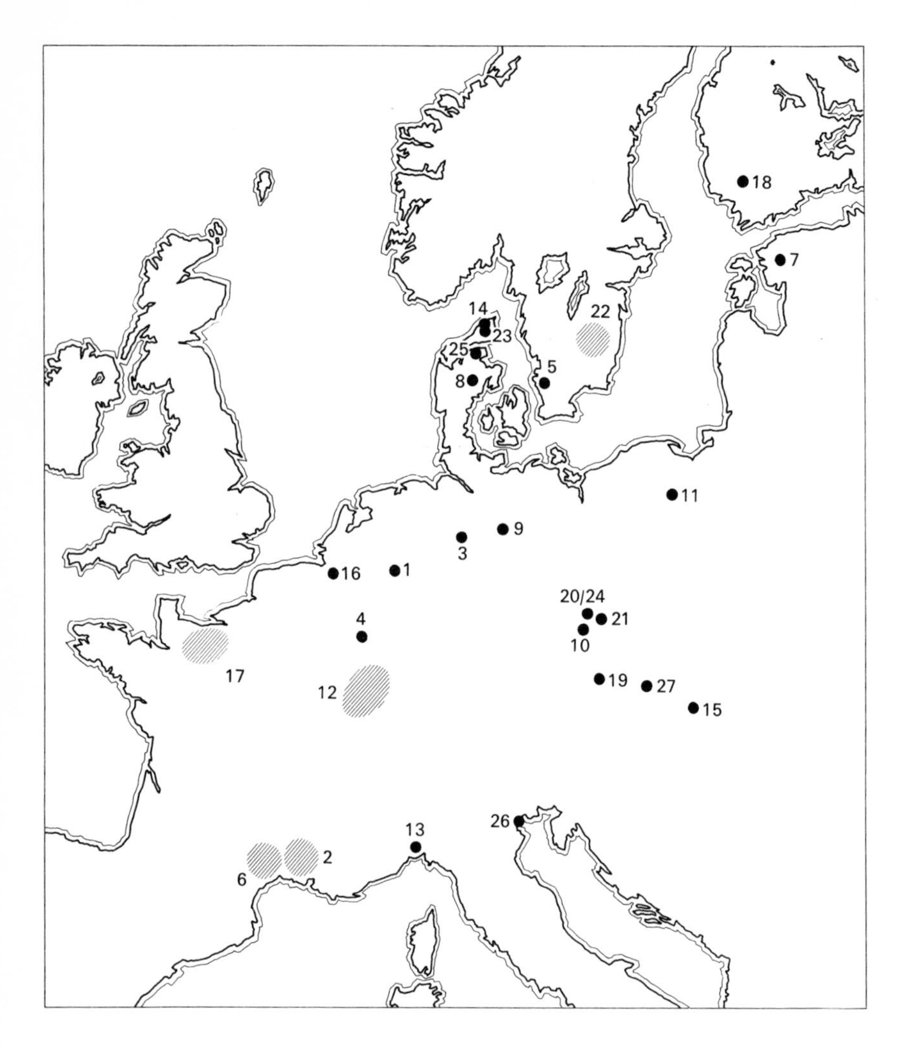

Major European Glassmaking Sites

1 Eigelstein, near Cologne, Germany
2 Gard, France
3 Grünenplan, Kreis Holzminden, Germany
4 Heindert, Canton d'Arlon, Luxembourg
5 Henrikstorp, in Skåne, Sweden
6 Herault, France
7 Húti (Dagö), Estonia
8 Hyttekaer, Tem Glarbo, Denmark
9 Junkersfelde , eastern Westphalia
10 Karlova Hut, Czechoslovakia
11 Kruszwica, Poland
12 Lorraine, about 40 sites
13 Monte Lecco, Apennines
14 Nejsum, in Vendsyssel, Denmark
15 Nitra, Slovakia
16 Nethen, Brabant, Belgium
17 Normandy, about 70 sites
18 Nötsjö, Finland
19 Počatky, Pelhrimov district, Czechoslovakia
20 Rejdice, Czechoslovakia
21 Sklenarice, Semily district, Czechoslovakia
22 Småland. Sweden (including Trestenshult)
23 Stenhule, in Tem Ǵarbo, Denmark
24 Syriste, Czechoslovakia
25 Tinsholt, south of Aalborg, Denmark
26 Venice, Italy (including Torcello)
27 Ververi Bityska, Czechoslovakia

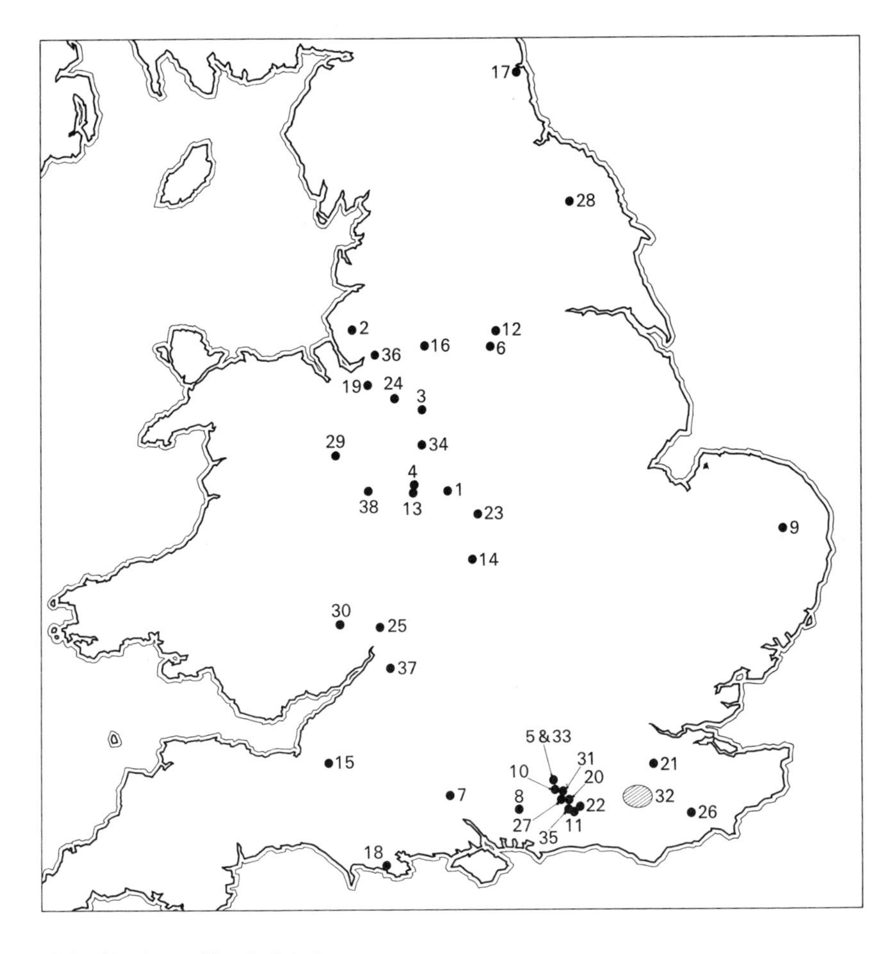

Major Glasshouse Sites in Britain

1 Bagot's Park, Staffordshire (OS 087 272) (14 sites)
2 Bickerstaffe, Lancashire (SD 441 036)
3 Biddulph, Congleton, Staffordshire
4 Bishop's Wood, Eccleshall, Staffordshire (6 sites)
5 Blunden's Wood, Hambledon, Surrey (SU 974 374)
6 Bolsterstone, South Yorkshire
7 Buckholt, Hampshire (SU 290 323)
8 Buriton, Hampshire (SU 740 168; SU 742 173)
9 Caistor St Edmund, Norfolk
10 Chaleshurst, Chiddingfold, Surrey (SU 9480 3323; SU 9485 3328)
11 Fernfold, Wisborough Green, West Sussex (TQ 0477 3208)
12 Gawber, South Yorkshire (SE 327 076)
13 Glasshouse Farm, Staffordshire (nr Bishop's Wood)
14 Glasshouse Wood, Kenilworth, Warwickshire (SP3T 308 717; SP37 313 720)
15 Glastonbury, Somerset
16 Haughton Green, Denton, Greater Manchester (SJ 942 946)
17 Jarrow and Monkwearmouth, Tyne and Wear
18 Kimmeridge Bay, Dorset
19 Kingswood, Delamere, Cheshire
20 Knighton's, Alfold, Surrey (TQ 016 341)
21 Knole, Kent
22 Malham Ashfold, Wisborough Green, West Sussex (TQ 0560 3007)
23 Mancetter, Warwickshire (SP 326 967)
24 Middlewich, Cheshire
25 Newent, Gloucestershire
26 Northiam, East Sussex
27 Plaistow, Kirdford, West Sussex (*ca* 4 sites)
28 Rosedale and Hutton Common, North Yorkshire (SE 745 932; SE 706 883)
29 Ruyton XI Towns, Salop (SJ 396 229)
30 St Weonards, Hereford and Worcester
31 Sidney Wood, Alfold, Surrey (TQ 0220 3372)
32 The Weald (over 40 sites)
33 Vann Copse, Hambledon, Surrey (SU 9842 3772)
34 Waterhays Farm, Staffordshire (SJ82 815 967)
35 Wephurst, Kirdford, West Sussex (TQ 0242 2938)
36 Wilderspool, Cheshire
37 Woodchester, Gloucestershire
38 Wroxeter, Salop

SITES TO VISIT

Museums

Fine glass vessel collections can be found either on show or in the stores of most large museums, notably the Victoria & Albert Museum, the British Museum and the Museum of London. Exceptional glass displays are also exhibited at the Cecil Higgins Art Gallery, Castle Close, Bedford; the Ashmolean Museum, Beaumont Street, Oxford; the Fitzwilliam Museum, Trumpington Street, Cambridge; the Royal Scottish Museum, Chambers Street, Edinburgh; the City Museum and Art Gallery, Queens Road, Bristol; the Castle Museum, Nottingham. The Cinzano Glass Collection, has been formed by Cinzano (U.K.) Ltd., Buckingham Gate, London.

There is an important display of mainly nineteenth and early twentieth-century glass at the Glass Museum in Moor Street, Brierley Hill, which has been removed for temporary display to the Museum and Art Gallery of Dudley, in St James's Road. Plans are under consideration for bringing together the Brierley Hill collection and the Stourbridge Glass collection, presently situated at the Council House, Mary Stevens Park, Stourbridge. It is proposed to move the two collections to what will become the Broadfield House Glass Museum, Kingswinford, near Brierley Hill.

The Haworth Art Gallery, Accrington, in the north west of England, houses the largest collection in Europe of decorative Art Nouveau glass made in the workshops of the famous American designer, Louis Comfort Tiffany (1848–1933).

Stained glass collections can be seen at The Stained Glass Museum at Ely Cathedral, the Victoria and Albert Museum, London, and the Burrell Collection, presently housed by Glasgow Museums and Art Galleries, but due to be resited in a new building in the grounds of Pollock House, Glasgow. Merseyside County Museums have a good stained glass collection at Speke Hall, Speke. An effort to list other museums with small stained glass collections is being made by Francis Skeat in a series of articles for the *Journal of the British Society of Master Glass Painters.*

For the increasing number of glass bottle enthusiasts, a fine collection of wine bottles, with some bottle moulds, can be seen at the Harvey's Wine

Museum, Denmark Street, Bristol. An important find of soft drink bottles with bottle equipment and gas making plant in the works of J. B. Bowler of Corn Street, Bath, is now being preserved by a trust.

Excavated material from British glasshouses is held in many museums, including the Victoria and Albert, the Pilkington Glass Museum, the Haslemere, Guildford, Littlehampton and Lewes museums. Glass from Rosemary Cramp's excavations at Jarrow and Monkwearmouth is housed in the new Bede Monastery Museum, Jarrow Hall, Church Bank, Jarrow, Tyne and Wear.

A gallery of glass technology was opened in 1968 at The Science Museum, South Kensington. Its displays concentrate almost entirely on manufacturing techniques, with examples of various products to illustrate the advances in glass technology.

Ryedale Folk Museum

A British reconstruction of a late sixteenth-century wood-fired furnace can be seen at the Ryedale Folk Museum, Hutton, North Yorkshire, built up on the remains of the Rosedale glasshouse.

Colonial National Historical Park, U.S.A.

Visitors to the Colonial National Historical Park, Jamestown, Virginia, U.S.A., can see the remarkable conjectural reconstruction of the wood-fired glasshouse which was in operation there in the early 1600s.

Pilkington Glass Museum

Although the Blunden's Wood glasshouse site was swept away by bulldozers after Eric Wood's excavation in 1960, a reconstruction of a medieval glasshouse based on this c 1330 site is on view at the Pilkington Glass Museum. The museum also displays a large model of a sixteenth-century north European wood-fired furnace with replicas of tools, and cullet from the Ruyton-XI-Towns glasshouse excavation (*pl 50*).

There is also a collection of material excavated from British sixteenth and seventeenth-century glasshouses in the museum's store.

The only 16 mm colour film of a glass furnace excavation, which takes in four seasons at the Haughton Green, Denton site of an early seventeenth-century coal-fired furnace, is held by the Pilkington Glass Museum. Although Pilkington Brothers found it economically impossible to go ahead with the reconstruction of the Haughton Green furnace, the remains of the furnace are still available for examination in the museum's store. A model of an early reverberatory furnace and a diorama of the later cone glasshouses also can be seen in the museum.

An example of the first commercially successful bottle making machine, invented by Michael J. Owens, dating to about 1907, is preserved in the

Pilkington Glass Museum's store. The machine, which came from United Glass Ltd, consists of six 'heads', each of which is a complete unit, containing a blank mould, a finishing mould and a neck-ring mould. During operation, the heads rotated round a central column and glass was drawn into the blank mould by suction from a revolving pot containing molten glass, the depth of which was kept constant. As each blank mould came over the pot, the machine dipped in order to lower the mould into the glass, and the pot rotated in order to remove the chilled glass. A knife then severed the glass hanging from the mould, and on the further rotation of the machine the blank mould opened, leaving a 'paraison' of glass hanging from the neck mould. The finishing mould then closed round the paraison and the bottle was blown to shape by compressed air. After the glass had been sufficiently chilled, the mould halves opened and the bottle was discharged, generally on to a conveyor, which took it to the lehr to be annealed.

The advances in glass technology in the nineteenth and twentieth centuries can best be seen in the superb displays, including scale models, at the Pilkington Glass Museum. The exhibits on the lower floor of the museum

Plate 50 Reconstruction of a typical sixteenth-century wood-fired rectangular furnace with replicas of tools. The cullet (bottom right) is from Ruyton-XI-Towns, Salop.

include flat glass making with the Bicheroux process,* twin grinding and polishing, and an up-to-date model of float glass manufacture. Glass in building, transport, decoration and science and technology is covered, plus displays on optical glass, fibreglass and safety glass, and a working laser with holograms of glass.

Surviving glasshouses

The best actual survival of a forest glass furnace is the *Bishop's Wood glasshouse*, Eccleshall, Staffordshire, found by T. Pape in 1931. It was one of six known furnace sites spread over three or four square miles, the other five having been destroyed. It was probably worked between *ca* 1584 and 1604.

A number of examples of the distinctive British cone glasshouses of the eighteenth and nineteenth centuries has survived. Complete cones can be seen at Catcliffe, near Sheffield; Lemington, near Newcastle-on-Tyne; Alloa, Clackmannan and the Red House cone in Wordsley, Stourbridge. Truncated cones may still be seen at Pilkington (Sheet Works), St Helens; Prewett Street, Bristol; the Dial Glass Cone, Amblecote; the White House Cone at Wordsley, Stourbridge and the Coalbournbrook Cone, Amblecote. Regrettably, a complete cone which belonged to the nineteenth-century Old Swan Glassworks, Liverpool, was pulled down recently (*pl 51*).

In 1740 William Fenny and a workman called Chatterton left the Bolsterstone glasshouse in South Yorkshire to start production at *Catcliffe*, ten and a half miles away. One of the two cones at Catcliffe, dating from about 1740, is claimed to be the oldest surviving example of this type of structure in Europe (*pl 52*). It is brick built to a height of sixty feet (18.2 m) with a base diameter of some forty feet (12.1 m). A small excavation by the City Museum, Sheffield, in 1962, cleared part of the furnace flue and the foundations of the pot furnace, and traced an extension of the furnace flue for fifty feet (15.2 m), as well as locating the walls of a service building to the west of the cone. The excavation established, from finds of cullet in the flue beneath the cone, that glassmaking was carried on there until at least 1900. Catcliffe was taken over by the May family in 1759, followed by Blunn & Booth in about 1833, when it produced flint glass and bottles. Documented Catcliffe glass from the third quarter of the nineteenth century includes jugs, vases and flasks decorated with opaque-white stripes. C. Wilcocks & Co. were glass bottle manufacturers at Catcliffe in 1901, the last known date when the glasshouse was in use.[1]

The cone at *Lemington*, near Newcastle-on-Tyne, is a vivid reminder of the great glassmaking past of the area, largely brought about by Sir Robert Mansell's search for cheap coal fuel in the early seventeenth century. The Lemington Glassworks was founded in 1787 by a group of gentlemen who

* The Bicheroux process was introduced in the early 1920s. Hot glass was poured from large pots between a pair of metal rollers, producing a flatter plate which needed less grinding than glass produced by the usual casting process.

Plate 51 Nineteenth-century cone glasshouse, Old Swan, Liverpool, which was recently demolished.

went into the glass trade under the name of the Northumberland Glass Company. Within a short time they had built four glasshouses on the site they had leased at Lemington, one of which was very lofty and brick built, this being the cone which can be seen today. The site is owned by Glass Tubes & Components Ltd., but the cone is no longer used.[2]

A handsome example of a glass cone is still preserved at the *Alloa* Glass Works, Clackmannan, Scotland, now owned by United Glass Ltd (*pl 53*). The cone was built in 1824 and is now referred to as '83 shop' at the works. According to a 1793 description of cone construction provided by the firm, the cone rose 'from a large circular foundation, several feet thick below-ground. On this foundation arches were built, and from fifteen feet above floor level, the structure was only nine inches thick. The inside diameter was

Plate 52 Catcliffe glasshouse, South Yorkshire, built *ca* 1740 with a 60 foot (18.46 metre) high cone.

about fifty feet on the ground and the height about ninety feet. From the foundation upward, the building leaned towards the centre, and through the opening at the top the smoke escaped.'

The most notable landmark in the Stourbridge district is the *Red House* cone, standing one hundred feet (30.5 m) high within the grounds of Stuart and Sons, makers of Stuart Crystal glass at Wordsley (*pl 54*). Judging by its layout, the Red House was probably erected after the cutting of the adjacent canal, which occurred within a year or two of the passing of the Stourbridge Canal Navigation Act, 1776. A document in the possession of Stuart and Sons records the sale of land on which the Red House now stands to a glass manufacturer, Richard Bradley, in 1788, but it has not been ascertained whether a glassworks was already standing or whether he bought the site with the intention of building a works.

The Red House Glassworks was run by the Bradleys, the Holts and then the Wainwrights, followed in 1828 by Richard Bradley Ensell II. He sold it to Rufford and Co. and it was leased in 1834 by Hodgetts and Davies, sold to William and Edward Webb in 1852, who leased it to William Hodgetts, then to Philip Pargeter. The firm was known as Mills Webb and Stuart, and later as Stuart and Mills. The Red House was leased to Frederick Stuart in 1881 and to Stuart and Sons in 1885. The cone was in continual use until 1939, since when it has been used as a storeroom. A preservation order was placed on it in 1966.

Stuart & Sons Ltd. expanded its business in 1915 with the acquisition of the *White House* Glassworks, which stands on the other side of the

Plate 53 The Alloa cone glasshouse, Clackmannan, Scotland, referred to by United Glass Ltd as '83 Shop'

Plate 54 The Red House cone glasshouse, which stands 100 feet (30.5 metres) high in the grounds of Stuart and Sons Ltd, Wordsley, Stourbridge.

Stourbridge-Wolverhampton main road. As with the Red House, the date of the erection of the White House cone, of which the bottom half can still be seen, is uncertain, but it was probably about 1812. The glassworks was closed on 9 December, 1935, when the old furnace collapsed, and in 1939 the cone was 'topped'.

Not too far from Wordsley, another truncated glass cone can be seen at the old *Dial* glassworks at the bottom of Stewkins Lane, Audnam, Amblecote. It is now owned by the glass manufacturers, Plowden & Thompson and used as a storeroom (*pl 55*). The date stone on the cone, which is situated on

Plate 55 Truncated glasshouse cone at the old Dial Glassworks, Audnam, Amblecote, used as a storeroom by its present owners. A datestone reading '1788' is situated below the roof line to the right of the nearest pair of roof supports.

the brickwork, just below the roof-line, says '1788', but the glasshouse was probably established in 1704.

A further partially cut down cone exists at Webb Corbett's glasshouse in Stourbridge. This was the former *Coalbournbrook* Cone in Stourbridge Road, Amblecote.[3]

The great glassmaking centre of Bristol has yielded few industrial remains, apart from the lower half of a glass cone in *Prewett Street*, which was converted into a restaurant in 1971. The Curator of Technology at Bristol's City Museum and Art Gallery, P. W. Elkin, put the date of the cone, which was formerly part of Prewett's glassworks, at about 1780.

1. *Lewis 1964*
2. *Cossons 1975*, 237
3. *Woodward 1976*

SOCIETIES AND PUBLICATIONS

General information on glassmaking and present-day manufacture can be obtained from the Glass Manufacturers Federation, 19 Portland Place, London, W1N 4BH, which also houses a good library on glassmaking. Other sources of information on glassmaking are the British Glass Industry Research Association (BGIRA), Northumberland Road, Sheffield, S1O 2UA; the Society of Glass Technology, 'Thornton', Hallam Gate Road, Sheffield, S1O 5BT and the joint Library of Glass Technology, Department of Ceramics, Glass and Polymers, University of Sheffield, Northumberland Road, Sheffield, S1O 2TZ. Excellent libraries on glassmaking exist at the Pilkington Glass Museum and Pilkington Research and Development Laboratories, Lathom, West Lancashire, and at the Corning Museum of Glass, Corning, New York, 14830, U.S.A. Book lists on most aspects of glassmaking may be obtained from both glass museums on request.

The Circle of Glass Collectors, or the Glass Circle, which was founded in 1937, is a society which embraces most amateur collectors, students and lovers of old glass in this country, and many from abroad. It produces much valuable information in the form of privately printed papers which are read at the society's monthly meetings from October to May, and which may be bought by the public.

GLOSSARY OF GLASSMAKING TERMS

Glassmaking terms are comprehensively listed and explained in *Glossary of Terms used in the Glass Industry* (British Standard 3447: 1962) and *Standard Definition of Terms Relating to Glass and Glass Products (American Society of Testing Materials* C162–71) but the short list of technical terms which follows should give enough initial guidance for archaeologists.

Applied decoration	Decoration added to a vessel in the form of trails, blobs and prunts
Blowing iron	Hollow iron rod used for blowing glass
Calcar	Kiln for the first fusion or fritting of raw materials
Chair	Glass-blower's seat
Chair or shop	Group of glsss-blowers working in a team
Cire perdue	The 'lost wax' process. The desired shape of the finished object would be carved in wax and a mould of refractory clay formed around this. The wax was then melted away and molten glass was cast into the mould. More likely, a model could have been made of an object in a material such as terracotta, which was then coated in wax and an outer mould put on top of that. When the whole was heated, the wax would melt and hot glass would be cast into the aperture, as in the first method. In a simpler method, powdered glass could be laid between two parts of a closed mould and fused to form the vessel.
Crisselling (crizzling)	Deterioration in the metal in the form of a network of tiny cracks, due to an excess of alkali
Crown glass	(see p 59)
Crucible (Pot)	Glassmaking pot made of fireclay
Cullet	Broken glass, glass waste

Cylinder glass	(see p 60)
Folded foot	Turned-over edge of the foot of a wine glass, resembling a hem
Frit	Partially fused ingredients, ready for final melt
Gaffer	Master glass-blower, head of the chair
Gather, gob	Molten glass taken from the furnace on a blowing iron or pontil
Glasshouse	Glassmaking factory
Grog	Broken pieces of crucible used in making new pots
Knop	A glass ball, hollow or solid, forming part of the stem of a vessel
Lehr, leer	Annealing kiln or tunnel
Marver	Slab of stone or iron upon which the hot glass is rolled after being gathered
Metal	Glass, either molten or cold
Milled edge	Glass transversely grooved on its edge, resembling the edge of a coin
Mix	Mixture of raw ingredients for glass
Mould-blown glass	Glass decorated with moulded patterns, such as diamonds or ribs, produced by blowing a glass bubble into a mould
Muff	see Cylinder
Pontil, puntee	Solid iron rod used to hold the glass while it is being worked
Pontil knock-off	Waste glass knocked from the end of a pontil, roughly resembling part of a bottle neck
Pontil mark	A rough circular mark on the base of a vessel, where it was fixed to the pontil
Pushed-in foot	A vessel foot of double thickness, where the vessel base has been pushed in on itself
Prunt	A seal of glass applied to the vessel, which may be either plain or moulded
Shearings, clippings	Waste glass which has fallen from the glassmaker's shears as he trimmed the edge of a vessel
Siege	Platform of clay or stone on which the crucibles rest in the furnace
Weathered glass	Glass with an opaque or iridescent film on its surface due to deterioration from weathering.

HOUGHTON LETTERS FOR THE IMPROVEMENT OF COMMERCE AND TRADE FIRST PUBLISHED 1683

List of Glass Houses in England and Wales
May 15, 1696. No. cxcviij.

An Account of all the Glass Houses in England & Wales	The several Counties they are in	The Number of Houses	And the sorts of Glass each House makes
In and about London and Southwark		9	For bottles
		2	Looking glass plates
		4	Crown glass and plates
		9	Flint glass and ordinary
Woolwich	Kent	1	Crown glass and plates
		1	Flint glass and ordinary
Isle of Wight	Hampshire	1	Flint glass and ordinary
Topsham nr Exon	Devonshire	1	Bottles
Odd Down nr Bath Chellwood	Somersetshire	1	Bottles
		1	Window glass
In and about Bristol		5	Bottles
		1	Bottles and window glass
		3	Flint glass and ordinary
Gloucester Newnham	Gloucestershire	3	Bottles
		2	Bottles Houses
Swansea in Wales	Glamorgan	1	Bottles
Oaken Gate	Shropshire	1	Bottles and window glass
Worcester	Worcestershire	1	Flint, green, and ordinary
Coventry	Warwickshire	1	Flint, green, and ordinary
Stourbridge	Worcestershire	7	Window glass
		5	Bottles
		5	Flint, green, and ordinary
Near Liverpool Warrington	Lancashire	1	Flint, green, and ordinary
		1	Window glass

Place	County	No.	Type
Nottingham	Nottingham	1	Bottles
Awsworth		1	Flint, green, and ordinary
Custom More		1	Bottles
Nr Awsworth		1	Flint, green, and ordinary
Nr Silkstone	Yorkshire	1	Bottles
Nr Ferrybridge		1	Bottles
		1	Flint, green, and ordinary
King's Lynn		1	Bottles
Yarmouth		1	Flint, green, and ordinary
		1	Bottles
Newcastle-upon Tyne	Northumberland	6	Window glass
		4	Bottles
		1	Flint, green, and ordinary
		88	

LOCAL CLAY ANALYSES

	BLUNDEN'S WOOD ca 1330	KNIGHTONS, ALFOLD ca 1550
	%	%
Silicon Dioxide	73.28	54.3
Titanium Dioxide	0.60	1.0
Aluminium Oxide	10.46	24.3
Ferric Oxide	5.92	6.9
Calcium Oxide	0.16	0.3
Magnesium Oxide	0.31	0.8
Titanium Dioxide+ Calcium Oxide+ Magnesium Oxide		
Potassium Oxide	1.00	2.6
Sodium Oxide	0.26	0.2
Zirconium Dioxide		Trace
Loss of Ignition	7.84	9.5

1. Quoted by the Geological Survey and Museum

CRUCIBLE ANALYSES

	BLUNDEN'S WOOD ca 1330	KNIGHTONS, ALFOLD ca 1550
	%	%
Silicon Dioxide	77.74	72.6
Titanium Dioxide	0.50	0.8
Aluminium Oxide	16.88	18.6
Ferric Oxide	2.26	2.2
Calcium Oxide	0.36	1.8
Magnesium Oxide	0.59	0.9
Potassium Oxide	1.45	2.3
Sulphur Trioxide		
Sodium Oxide	0.12	Trace
Loss on Ignition	0.04	0.7

ST WEONARD'S late 16th/early 17th centuries Forest of Dean Clay[1]	ST WEONARD'S late 16th/early 17th centuries Stour-bridge Clay	HAUGHTON GREEN ca 1605–53
%	%	%
62.00	68.58	64.5
	1.75	0.8
19.00	25.21	19.3
3.00	2.52	3.4
	0.33	0.1
	0.28	0.1
5.78		
	0.96	2.6
	0.37	0.7
10.22	8.31	8.3

BICKER-STAFFE ca 1600	ST WEONARD'S late 16th/ early 17th centuries	HAUGHTON GREEN ca 1605–53	GAWBER PHASE I ca 1690–1735
%	%	%	%
77.1	75.93	77.0	78.6
1.0	1.05	1.3	
18.8	20.06	17.4	17.0
0.91	1.25	1.1	1.3
0.1	0	0.6	
0.4	0	0.3	
0.9	1.22	1.3	1.1
Trace		0.2	
	0.17	0.9	0.2
0.4			

GLASS ANALYSES

	BLUNDEN'S WOOD ca 1330	KNIGHTONS ALFOLD ca 1550	HUTTON late 16th century	ROSEDALE late 16th century
	%	%	%	%
Silicon Dioxide	57.00	74.8	58.98	58.84
Lead Oxide				
Aluminium Oxide +Titanium Dioxide			4.98	5.81
Titanium Dioxide	0.08	0.5		
Aluminium Oxide	4.78*	6.1		
Ferric Oxide	1.32	2.7	1.51	1.57
Manganous Oxide			0.37	0.30
Calcium Oxide	17.50	0.9	24.59	20.15
Magnesium Oxide	6.95	0.5	2.28	3.25
Potassium Oxide	9.00	11.8	5.16	6.20
Phosphorus Pentoxide				
Manganese Oxide		0.3		
Barium Oxide		0.2		
Sodium Oxide	3.40	1.9	1.84	3.54
Lithium Oxide				
Sulphur Trioxide		Trace		
Loss on Ignition	0.08			

*With trace of manganese up to 0.2%

BICKER- STAFFE ca 1600	ST WEONARD'S late 16th/early 17th centuries	JAMESTOWN early 17th century	HAUGHTON GREEN ca 1605–53	GAWBER PHASE I ca 1690–1735	GAWBER PHASE II 18th/19th centuries
%	%	%	%	%	%
60.4	58.00	59.5	57.8	53.0	56.5
				34.4	
0.1			0.2		0.3
2.3			7.0	1.5	5.5
0.61	1.4	1.6	1.2	0.18	2.3
19.5	19.6	21.5	20.00	0.5	22.00
4.7	3.9	3.8	4.8	0.1	4.9
1.4	9.00	3.9	1.8	7.5	2.5
3.4					2.0
0.8			1.1	n.d.	1.2
0.2			0.1		0.1
5.9	2.00	1.4	5.8	0.3	2.2
					0.06
0.5			0.2		0.3

BIBLIOGRAPHY

All publications first appeared in London unless otherwise stated.

CHAPTER ONE What is Glass?

DILLON, E. 1907: *Glass*
DUNCAN, G. S. 1960: *Bibliography of Glass* (Sheffield)
ELVILLE, E. M. 1961: *The Collector's Dictionary of Glass*
HARDEN, D. B., PAINTER, K. S., PINDER-WILSON, R. H. and TAIT, H. 1968: *Masterpieces of Glass*
HAYNES, E. Barrington 1948, 1970: *Glass Through the Ages*
HONEY, W. B. 1946: *Glass: A Handbook*
KÄMPFER, F. and BEYER, K. G. 1966: *Glass: A World History*
MALONEY, F. J. T. 1967: *Glass In the Modern World, A Study in Materials Development*
MARIACHER, G. 1954: *L'Arte del Vetro*. (Verona)
SCHLOSSER, I. 1956: *Das Alte Glas* (Braunschweig)
SCHMIDT, R. 1922: *Das Glas* (2nd edition, Berlin and Leipzig)
VÁVRA, J. R. 1954: *5000 Years of Glassmaking* (Prague)
VOSE, R. Hurst 1975: *Glass* (Connoisseur Illustrated Guides)
WEISS, G. 1971: *The Book of Glass*

CHAPTER TWO The Origins of Glassmaking

ABRAMIĆ, M. 1959: 'Eine Römische Lampe mit Darstellung des Glasblasens' *Bonner Jahrbücher* CLIX, 149ff
BARAG, D. 1970–71: 'Glass Pilgrim Vessels from Jerusalem', Parts I, II and III. *Journal of Glass Studies* XII, 35–63, XIII, 45–63
BRILL, R. H. 1963: 'Ancient Glass', *Scientific American* vol. 209, no. 5, 120–130
CALEY, E. R. 1962: *Analyses of Ancient Glasses, 1790–1957* (Corning, New York)
CHARLESTON, R. J. 1963: 'Glass "Cakes" as Raw Material and Articles of Commerce', *Journal of Glass Studies* V, 54–67.

CHARLESTON, R. J. 1964: 'Wheel-Engraving and Cutting: Some Early Equipment', *Journal of Glass Studies* VI, 83–101

DOPPELFELD, C. 1965: 'Die Kölner Glasöfen vom Eigelstein' *International Commission on Glass conference*, Brussels, paper 236

ENGLE, A. 1978: *Ancient Glass in its Context No. 10 An Illustrated Companion to Readings in Glass History Nos. 1–8* (Jerusalem)

FARNSWORTH, M. and RITCHIE, P. D. 1938; 'Spectrographic studies in ancient glass. Egyptian glass, mainly of the 18th Dynasty, with special reference to its cobalt content', *Technical Studies* VI, no. 3

GARNER, H. 1956: 'An Early Piece of Glass from Eridu' *Iraq* XVIII, 147–149

HARDEN, D. B. 1936: *Roman Glass from Karanis* (Ann Arbor)

HARDEN, D. B. 1968: 'Pre-Roman Glass', *See* Harden, Painter, Pinder-Wilson and Tait (1968), 11–35

ISINGS, C. 1957: *Roman Glass from Dated Finds* (Groningen)

LABINO, D. 1966: 'The Egyptian Sand-Core Technique: A New Interpretation', *Journal of Glass Studies* VIII, 124–127

LAMM, C. J. 1928: *Das Glas von Samarra* (Berlin)

NEUBURG, F. 1962: *Ancient Glass*

NEWTON, R. C. 1978: 'Colouring agents used by medieval glassmakers', *Glass Technology* xix, no. 3, 59–60

OLIVER, A. 1968: 'Millefiori Glass in Classical Antiquity', *Journal of Glass Studies* X, 48–70

OPPENHEIM, A. L., BRILL, R. H., BARAG, D. and SALDERN A. von 1970: *Glass and Glassmaking in Ancient Mesopotamia* (The Corning Museum of Glass Monographs III, Corning, New York)

PAINTER, K. S. 1968: 'Roman Glass'. *See* Harden, Painter, Pinder-Wilson and Tait (1968), 36–90

PETRIE, W. M. Flinders 1894: *Tell-el-Amarna*

PINDER-WILSON, R. H. 1968: 'Pre-Islamic Persian and Mesopotamian, Islamic and Chinese'. *See* Harden, Painter, Pinder-Wilson and Tait (1968), 98–126

SALDERN, A. von 1959: 'Glass Finds at Gordion', *Journal of Glass Studies* I, 23–49

SALDERN, A. von 1966: 'Mosaic Glass from Hasanlu, Marlik and Tell al-Rimah', *Journal of Glass Studies* VIII, 9–25

SCHÄFER, F. and ZECCHIN, L. 1968: 'Two Pragmatic Views on Vasa Dietreta I, II' *Journal of Glass Studies* X, 176–179

SCHÜLËR, F. 1959: 'Ancient Glassmaking Techniques: The Moulding Process', *Archaeology* XII, no. 1, 47–52

SMITH, R. W. 1964: 'History revealed in ancient glass' *National Geographic Magazine* CXXVI, 3, 346–369

TURNER, W. E. S. 1954, 1956, 1959: 'Studies in Ancient Glasses and Glass-making Processes' Parts I-VI, *Journal of the Society of Glass Technology*

WEINBERG, G. D. 1968: 'Roman Glass Factories in Galilee', *Bulletin*, Museum Haaretz X, 49–50

CHAPTER THREE Glassmaking on the Continent: Middle Ages to 1700

AGRICOLA, G. 1556: *De re metallica,* trans. and ed. by H. C. Hoover and L. H. Hoover (New York 1950)

BIRINGUCCIO, V. 1540: *De la Pirotechnia Libri X dove ampiamente si tratta . . . di ogni sorte et diversita di miniere.* (Venice). English edition by C. S. Smith and M. T. Gnudi, (New York 1943)

BARRELET, J. 1953: *La Verrerie en France de l'époque Gallo-Romaine à nos jours* (Paris)

CHARLESTON, R. J. and ANGUS-BUTTERWORTH, L. M. 1957: 'Glass' in Singer, C., Holmyard, E. J., Hall, A. R., and Williams T. I. 1957: *A History of Technology*, 206–244 (Oxford)

CHARLESTON, R. J. 1962: 'Some tools of the glassmaker in medieval and renaissance times with special reference to the glassmaker's chair'. *Glass Technology* III, 107–112

CHARLESTON, R. J. 1972: 'Enamelling and Gilding on Glass' *The Glass Circle I* (Newcastle-upon-Tyne), 18–32

CHARLESTON, R. J. 1978: 'I. Glass Furnaces through the Ages. II. A Gold and Enamel Box in the Form of a Glass Furnace', *Journal of Glass Studies* XX, 9–44

CORNING MUSEUM OF GLASS 1958: *Three Great Centuries of Venetian Glass* (New York)

EGG, E. 1962: *Die Glashütten zu Hall und Innsbruck im 16 Jahrhundert* (Innsbruck)

FROTHINGHAM, A. W. 1963: *Spanish Glass*

GASPARETTO, A. 1958: *Il vetro di Murano dalle origini ad oggi* (Venice)

GASPARETTO, A. 1965: 'Les Fouilles de Torcello et leur apport à l'histoire de la verrerie de la Vénétie dans le Haut Moyen Age'. *International Commission on Glass*, Brussels, Paper 239

HEJDOVÁ, D. 1963: 'A medieval Glass Works at Sklenarice in North Bohemia', *Czechoslovak Glass Review* XVIII, No. 12, 360–364

HEJDOVÁ, D. 1975: 'Types of Medieval Glass Vessels in Bohemia', *Journal of Glass Studies* XVII, 142–150

HETTEŠ, K. 1960: *Venezianisches Glas* (Prague)

KUNCKEL, J. 1679: *Ars Vitraria Experimentalis* (Frankfurt and Leipzig)

LADAIQUE, G. 1957: 'L'industrie du verre et du cristal, dans les départements de Meurthe-et-Moselle et des Vosges'. *Annales de L'Est*, no. 2, 125–144 (Nancy)

MANNONI, T. 1972: 'A Medieval Glasshouse in the Genoese Apennines, Italy', *Medieval Archaeology* XVI, 143–145

MOORHOUSE, S. et.al. 1972: 'Medieval Distilling Apparatus of Glass and Pottery' *Medieval Archaeology* XVI, 79–121

NERI, A. 1612: *L'Arte Vetraria*, (Firenze) English edition by Christopher Merrett (1662)

POLAK, A. 1975: *Glass – Its Makers and its Public*

ROOSMA, M. 1966: *Glassworks of Húti*, (Tallinn, Estonia)

SEITZ, H. c. 1939: 'Factorier och Manufakturer . . .', *Svenska Folket Genom Tiderna* Malmö, 279–282

TABACZYŃSKA, E. 1970: 'Remarks on the Origin of the Venetian Glassmaking Centre' (Results of the Italo-Polish excavations at Torcello 1961–62) *Studies in Glass History and Design*, papers read to Committee B Sessions of 8th International Congress on Glass (1968) edited by R. J. Charleston, W. Evans and A. E. Werner

TAIT, H. 1979: *The Golden Age of Venetian Glass*

TERLINDEN, A. M. 1980: 'Post Medieval Glass-making in Belgium: The Excavation of a Seventeenth-Century Furnace at Nethen (Brabant)', *Post Medieval Archaeology* XIV

THEOPHILUS, P. c 1100–1150: *De Diversis Artibus* or *Schedula Diversarum Artium*, Translated by C. R. Dodwell (1961), also by J. C. Hawthorne and C. S. Smith, (Chicago 1963)

CHAPTER FOUR Glassmaking on the Continent: 1700–1900

BEARD, G. 1968: *Modern Glass*

BORSOS, B. 1963: *Glassmaking in Old Hungary* (Budapest)

BUCKLEY, W. 1926: *European Glass*

CHAMBON, R. 1955: *L'Histoire de la verrerie en Belgique du IIe siècle à nos jours* (Brussels)

CHARLESTON, R. J. 1965: 'Wheel-Engraving and Cutting: Some Early Equipment – II Water-Power and Cutting', *Journal of Glass Studies* VII, 41–55

DIDEROT, D. and D'ALEMBERT, J. 1751–65: *Encyclopédie ou dictionnaire raisonné des sciences, des arts et des métiers*, (Geneva).

DIDEROT, D. and D'ALEMBERT, J. 1762–72: *Receuil de planches sur les sciences, les arts libéraux, les arts méchaniques avec leur explication* (Paris 10 vols)

DIDEROT, D. 1959: *A Diderot Pictorial Encyclopaedia of Trades and Industry, II.* (New York), 209–275

DOUGLAS, R. W. and FRANK, S. 1972: *A History of Glassmaking* (Henley-on-Thames)

GARDNER, P. V. 1971: *The Glass of Frederick Carder* (New York)

HODKIN, F. W. and COUSEN, A. 1925: *A Textbook of Glass Technology* (New York)

HUME, I. N. 1976: 'Archaeological Excavations on the Site of John Frederick Amelung's New Bremen Glassmanufactory, 1962–1963', *Journal of Glass Studies* XVIII 137–214

JANNEAU, G. 1931: *Modern Glass*

KOCH, R. 1964: *Louis C. Tiffany, Rebel in Glass* (New York)

LARDNER, D. 1972: *The Manufacture of Porcelain and Glass* Reprint of 1832 edition

MCKEARIN, G. S. and H. 1941: *American Glass* (New York)

PAZAUREK, G. E. 1923: *Gläser der Empire und Biedermeierzeit* (Leipzig)

POLAK, A. 1953: *Gammelt norsk glass* (Old Norwegian glass) (Oslo)

GLASS

Polak, A. 1962: *Modern Glass*
Revi, A. C. 1959: *Nineteenth Century Glass – Its Genesis and Development* (New York, rev. ed. 1968)
Saldern, A. von 1965: *German Enamelled Glass* (New York)
Scoville, W. C. 1950: *Capitalism and French Glassmaking, 1640–1789.* (Berkeley and Los Angeles)
Seela, J. 1974: 'The Early Finnish Glass Industry', *Journal of Glass Studies* XVI, 57–86
Stennett-Wilson, R. 1958: *The Beauty of Modern Glass*
Tait, H. 1968: 'European: Middle Ages to 1862', See Harden, Painter, Pinder-Wilson and Tait, 127–192

CHAPTER FIVE Glassmaking in Britain

Alcock, L. 1963: *Dinas Powys*
Atkinson, D. 1929: 'Caistor Excavations', *Norfolk Archaeology* XXIV, 108ff
Buckley, F. 1925: *A History of Old English Glass*
Cramp, R. 1970: 'Decorated Window Glass and Millefiori from Monkwearmouth', *Ant. J.* L, 327–335
Cramp, R. 1975: 'Window Glass from the Monastic Site of Jarrow. Problems of Interpretation', *Journal of Glass Studies* XVII, 88–96
Engle, A. 1977: 'Jean Carré of Arras', *Readings in Glass History* VIII, 1–20
Godfrey, E. S. 1975: *The Development of English Glassmaking 1560–1640* (Oxford)
Haden, H. J. 1971: *The 'Stourbridge Glass' Industry in the 19 c* (A study of the Glass Industry in Stourbridge, Brierley Hill and Dudley) Tipton
Harden, D. B. 1956: 'Glass Vessels in Britain and Ireland, A.D. 400–1000', in D. B. Harden (ed.) *Dark Age Britain: studies presented to E. T. Leeds*
Harden, D. B. 1961: 'Domestic Window Glass: Roman, Saxon and Medieval' in Jope, E. M., (ed.) *Studies in Building History*, 39–63
Harden, D. B. 1968: In 'An Early Iron Age Burial at Welwyn Garden City' by I. M. Stead, *Archaeologia* CI pp. 14–16
Harden, D. B. 1978: 'Anglo-Saxon and Later Medieval Glass in Britain: Some Recent Developments' *Medieval Archaeology*, 1–24
Hartshorne, A. 1897: *Old English Glasses.* Re-published as *Antique Drinking Glasses: A Pictorial History of Glass Drinking Vessels* (New York, 1967)
Hughes, G. B. 1956: *English, Scottish and Irish Table Glass from the 16th Century to 1820*
Hughes, M. J. 1972: 'A Technical Study of Opaque Red Glass of the Iron Age in Britain', *Proceedings of the Prehistoric Society* XXXVIII, 98–107
Hume, I. N. 1961: 'The Glass Wine Bottle in Colonial Virginia', *Journal of Glass Studies* III, 91–119
Kenyon, G. H. 1967: *The Glass Industry of the Weald* (Leicester)
McKerrell, H. 1972: 'On the Origins of British faience beads and some aspects of the Wessex-Mycenae relationship'. *Proc. Prehist. Soc.* XXXVIII, 286–301

MORRIS, B. 1978: *Victorian Table Glass and Ornaments*
NEALE, F. 1974: Thesis on the topography of medieval Bristol, University of London. Refs. to domestic window glass in Bristol sent to author, September, 1974
NEWTON, R. G. and RENFREW C. 1970: 'British Faience Beads Reconsidered', *Antiquity* XLIV, 199–206
PELLATT, A. 1849: *Curiosities of Glassmaking*
POWELL, H. J. 1923: *Glass-Making in England* (Cambridge)
SALZMAN, L. F. 1952: *Building in England down to 1540; A Documentary History* (Oxford)
SIMMS, R. 1894: *Contributions towards a History of Glass Making and Glass Makers in Staffordshire with an extraordinary tale entitled 'A Legend of the Glasshouse' founded on fact, altered from the original by R. S.* (Wolverhampton)
THORPE, W. A. 1935: *English Glass* (3rd edition 1961)
WARREN, P. 1970: *Irish Glass*
WAKEFIELD, H. 1961: *Nineteenth Century British Glass*
WILSON, D. R. 1974: 'Roman Britain in 1973: I. Sites Explored; Wroxeter', *Britannia* V, 428–9

CHAPTER SIX Archaeological Evidence in Britain

ANON 1967: 'An Early Glass Manufactory at Ruyton-XI-Towns'. *Shropshire News Letter* No. 32, 2–3
ASHURST, D. 1970: 'Excavations at Gawber Glasshouse, near Barnsley, Yorkshire', *Journal of the Society for Post-Medieval Archaeology* IV, 92–140
ASHURST, D. 1969/1972: References to Bolsterstone Glasshouse excavation, *Post-Medieval Archaeology* (1969) p 207 and (1972) p 221
BOSWELL, P. G. H. 1918: *Memoir on British Resources of Sands and Rocks used in Glassmaking*
BOWIE, G. 1974: *Preliminary Report on Ballycastle Glass Kiln Excavation* (photostat)
BRIDGEWATER, N. P. 1963: 'Glasshouse Farm, St Weonards: A Small Glass-working Site', *Trans. Woolhope Nat. Hist. and Field Club* XXXVII, 300–315
CHARLESWORTH, D. 1967: 'A Primitive Glass Furnace in Cairo', *Journal of Glass Studies* IX, 129–132
CROSSLEY, D. W. 1967: 'Glassmaking in Bagot's Park, Staffordshire, in the Sixteenth Century', *Post-Medieval Archaeology* I, 44–83
CROSSLEY, D. W. and ABERG, F. A. 1972: 'Sixteenth Century Glass-Making in Yorkshire: Excavations at Furnaces at Hutton and Rosedale, North Riding, 1968–1971, *Post-Medieval Archaeology* VI, 107–159
DANIELS, J. S. 1950: *The Woodchester Glasshouse*, (Gloucester)
DREW, J. H. 1967–1970: 'A Glasshouse at Ashow, near Kenilworth, Warwickshire', *Transactions of the Birmingham and Warwickshire Archaeological Society* LXXXIV, 187–188

GLASS

HOGAN, D. E. 1968: 'The Du Houx and the Haughton Green Glasshouse', *Studies in Glass History and Design*. Papers read to Committee B Sessions of the Eighth International Congress on Glass, 24–26. Editors R. J. Charleston, W. Evans, A. E. Werner

HURST, R. 1968: 'The Bickerstaffe Glasshouse', *Studies in Glass History and Design*. Papers read to the Committee B Sessions of the Eighth International Congress on Glass, 26–30. Editors R. J. Charleston, W. Evans, A. E. Werner

MAY, T. 1904: *Warrington's Roman Remains* (Warrington) 37–53

NEF, J. U. 1932: *The Rise of the British Coal Industry* 2 vols

SMITH, R. S. 1962: 'Glassmaking at Wollaton in the Early 17th Century', *Trans. Thoroton Society of Nottingham* LXVI, 24–34

VOSE, R. Hurst 1971: Anon. *The Denton Glass Excavation. The results of the excavations at a unique glassmaking site* Exhibition Catalogue, North West Museum and Art Gallery Service and The Pilkington Glass Museum

VOSE, R. Hurst 1972: 'Bickerstaffe and Haughton Green Excavations', Extrait des *Annales du 5e Congrès de l'Association Internationale pour l'Histoire du Verre, Liège*

VOSE, R. Hurst 1977: 'Glassmaking at Kingswood, Delamere, Cheshire' in *Vale Royal Abbey and House* Winsford Local Historical Society and the Michaelmas Trust) 34–36

WESTROPP, M. S. Dudley 1920: *Irish Glass, An Account of Glass-making in Ireland from the 16th Century to the Present Day*

WHATMOOR, P. 1976: 'Philip Delves Into History In Kimmeridge Bay', *B.P. News* February, 5

WOOD, E. S. 1965: 'A Medieval Glasshouse at Blundens Wood, Hambledon, Surrey', *Surrey Archaeological Collections* LXII, 54–79

WOOD, E. S. (forthcoming): 'A Sixteenth Century Glasshouse: Knightons, Alfold, Surrey' *Surrey Archaeological Society Research Volumes*

CHAPTER SEVEN Glasshouse Excavation

AITKEN, M. J. 1974: *Physics and Archaeology* (Oxford)

ANDERSON, C. A. ed. 1973: *Microprobe Analysis*

BIEK, L. and BAYLEY, J. 1979; 'Glass and other vitreous materials', *World Archaeology* XI, no. 1, 1–25

BIMSON, M. and ORGAN, R. M. 1968: 'The safe storage of unstable glass', *Museum News* XLVI, 39–47

BRILL, R. H. 1961: 'The Record of Time in Weathered Glass', *Archaeology* XIV, no. 1, 18–22

BRILL, R. H. 1975: 'Crizzling – A Problem in Glass Conservation', *Conservation in Archaeology and the Applied Arts,* Stockholm Congress, 11.C, 121–134

BRILL, R. H.: *Recommendations on Cleaning of Glass Finds in the Field*, (New York, photostat)

BRILL, R. H., FLEISCHER, R. L., PRICE P. Butford and WALKER, R. M. 1964: 'The Fission–Track Dating of Man-Made Glasses: Preliminary Results', *Journal of Glass Studies* VI 151–157

BRILL, R. H. and HOOD, H. P. 1961: 'A New Method for Dating Ancient Glass', *Bulletin, Central Glass and Ceramic Institute* VIII, no. 1, 51–54

BRILL, R. H. and MOLL S. 1961: 'The Electron Beam Probe Microanalysis of Ancient Glass', *Proceedings of the International Institute for Conservation, Rome Conference* 260–276 (mimeographed)

CHIRNSIDE, R. C. and PROFFITT, P. M. C. 1963: 'The Rothschild Lycurgus Cup: An Analytical Investigation', *Journal of Glass Studies* V, 18–23

CORNING MUSEUM OF GLASS 1977: *The Corning Flood: Museum under Water*, (New York)

HARRINGTON, J. C. 1972: *A Tryal of Glasse. The Story of Glassmaking at Jamestown* (Virginia)

HUDSON, A. P. and NEWTON R. 1976: 'A means for the in-situ identification of medieval glass by the detection of its natural radioactivity', *Archaeometry* XVIII, 229–232

HUDSON, J. P. 1961: 'Seventeenth Century Glass Wine Bottles and Seals Excavated at Jamestown', *Journal of Glass Studies* III, 79–91

HURST, J. 1970: *The Rosedale Glass Furnace and the Elizabethan Glassworkers* (Ryedale)

LANFORD, W. A. 1977: 'Glass Hydration: A Method of Dating Glass Objects', *Science* CXCVI, 975–976

NEWTON, R. G. 1966: 'Some problems in the dating of ancient glass by counting the layers in the weathering crust', *Glass Technology* VII, 22–25

NEWTON, R. G. 1971: 'The enigma of the layered crusts on some weathered glasses, a chronological account of the investigations', *Archaeometry* XIII, 1–9

NEWTON, R. G. 1974: *The Deterioration and Conservation of Painted Glass: A Critical Bibliography and Three Research Papers*

NISHIMURA, S. 1971: 'Fission Track Dating on Archaeological Material from Japan', *Nature* CCXXX, 242–243

SHAW, G. 1965: 'Weathered Crusts on Ancient Glass', *New Scientist* XXVII, no. 454, 290–291

WEIER, L. E. 1973: 'The deterioration of inorganic materials under the sea', *Bull. Inst. Archaeol.* XI, 131–163

WERNER, A. E. 1966: 'The Care of Glass in Museums', *Museum News*, June, 45–49

WERNER, A. E. 1968: 'Analytical Methods in Archaeology', *Analytical Chemistry* XL, no. 2, 28A–42A

WOLSKY, S. P. and CZANDERNA, A. W. (eds.) 1975: *Methods of Surface Analysis. Methods and Phenomena 1* (New York)

GLASS

SITES TO VISIT

BARKER, T. C. 1977: *The Glassmakers. Pilkington: the rise of an international company 1826–1976*

CARVEL, J. L. 1953: *The Alloa Glass Work, An Account of its Development since 1750.* (Edinburgh)

COSSONS, N. 1975: *The B.P. Book of Industrial Archaeology* 231—241

HADEN, H. J. 1949: *Notes on the Stourbridge Glass Trade* (Brierley Hill)

HUDSON, K. 1971: *A Guide to the Industrial Archaeology of Europe*, (Madison, N.J.) (Glass factories and museums in Europe.)

LEWIS, G. D. 1964: *The South Yorkshire Glass Industry* (Sheffield)

STUART, W. E. C. 1971: *History of 'Stuart Crystal'* (photostat)

WOODWARD, H. W. 1976: 'Stourbridge Glass, Aspects of Making and Decorating', *Glass Circle*, January

WOODWARD, H. W. 1978: *'Art, Feat and Mystery' The Story of Thomas Webb & Sons, Glassmakers* (Stourbridge)

INDEX